Humans and Cyber Security

Cyber security incidents are often attributed to "human error". The discipline of human factors recognises the importance of identifying organisational root causes, rather than focusing on individual actions or behaviours. *Humans and Cyber Security: How Organisations Can Enhance Resilience through Human Factors* delivers an applied approach to capturing and mitigating the risk of the human element in cyber security and proposes that it is easier to change organisational practices than to change individual behaviour.

This book first identifies undesirable behaviours and practices, then analyses why they occur, and finally offers mitigating actions. Models of behavioural motivations will be discussed alongside individual vulnerabilities. Organisational vulnerabilities will form the main focus of an applied approach to capturing and mitigating the risk of the human element in cyber security. It concludes with recommended processes that involve talking to a range of individuals across the organisation. Backed up with practical materials to facilitate data collection, applied examples, and mitigating strategies to address known human vulnerabilities, this book offers the reader a complete view of understanding and preventing cyber security breaches.

The solutions in this book will appeal to students and professionals of human factors, security, informational technology, human resources, and business management.

Amanda Widdowson is the Head of Human Factors and User Experience Capability at a large, multinational organisation. She has been applying human factors for over 25 years, mostly in the Defence and Rail sectors. A former President and Chartered Fellow of the Chartered Institute of Ergonomics and Human Factors, she served as a trustee on the Executive Board and as Chair of the Honours Committee until 2022 before starting a second term on the board. In 2018, she received the CIEHF's Lifetime Achievement award for "significant contributions to the application of ergonomics/human factors". In 2019, she won a "Highly Commended" award for the Management

Consulting Association's Thought Leading Consultant of the Year and the UK Ministry of Defence's H Rowbotham award for "outstanding contribution to the field of Human Factors Integration". In 2024, she was presented with the CIEHF's annual Institute Lecture award and was invited to speak about human factors in cyber security in this keynote address.

Humans and Cyber Security

How Organisations Can Enhance Resilience Through Human Factors

by Amanda Widdowson

CRC Press
Taylor & Francis Group
Boca Raton London New York

CRC Press is an imprint of the
Taylor & Francis Group, an **informa** business

Designed cover image: © 2025 Shutterstock

First edition published 2025
by CRC Press
2385 NW Executive Center Drive, Suite 320, Boca Raton FL 33431

and by CRC Press
4 Park Square, Milton Park, Abingdon, Oxon, OX14 4RN

CRC Press is an imprint of Taylor & Francis Group, LLC

© 2025 Amanda Widdowson

ISBN: 978-1-032-54784-8 (hbk)
ISBN: 978-1-032-54831-9 (pbk)
ISBN: 978-1-003-42768-1 (ebk)

DOI: 10.1201/978-1-003-42768-1

Typeset in Times
by codeMantra

Contents

Introduction

1

Cyber security incidents are often attributed to "human error". The challenge of how to measure human vulnerabilities in cyber security can be addressed through human factors, a scientific discipline with a basis in psychology, physiology, and engineering. Human factors use knowledge of human mental and physical strengths and limitations to design equipment, systems, or organisations around people. The aim is to enhance efficiency, safety, and the user experience. Rather than focussing on individual actions or behaviours, it recognises the importance of identifying organisational root causes of incidents.

The costs of cybercrime and the benefits of applying established human factors knowledge and principles to understand and mitigate the risk of cyber security incidents will be discussed. Cyber-attack examples will further illustrate the relevance of human factors.

Models of behavioural motivations will be considered alongside individual vulnerabilities. However, organisational vulnerabilities will form the focus of an applied approach to capturing and mitigating the risk of the human element on cyber security. We all make errors, no matter how well trained we are. Therefore, it is not enough to simply raise awareness about cyber security. It is also necessary to *enable* good behaviours. From a practical point of view, it is much easier to change organisational practices than individual behaviour. To this end, solutions will be provided to enhance resilience. Practical materials will facilitate data collection, and applied examples will illustrate the approach and lessons learned.

This book is likely to be of interest to practitioners and students with an interest in cyber security and human factors.

DOI: 10.1201/9781003427681-1

Problem definition – why consider human factors in cyber security?

2

WHY CONSIDER CYBER SECURITY?

Before we can answer the question 'Why consider human factors in cyber security?', we first need to consider the importance of cyber security.

Cybercrime is expensive. Incidents typically affect access to files, software, networks, websites, and networks and compromise sensitive data. The estimated global cost of cybercrime is expected to reach 15.63 trillion US dollars in 2029.[1] In the European Union (EU) and United Kingdom (UK), fines for breach of General Data Protection Regulations (GDPR)[2] can reach £17.5 million or 4% of an organisation's annual turnover (whichever is greater), (ICO guide).[3] Governments are responding with significant investment in cyber security. For example, the UK proposed £2.6 million to address cyber security and legacy Information Technology (IT) from 2022 to 2025[4] and the US National Cyber Security Strategy[5] advocates investment in research and development. Both the UK and US strategies aim to place more responsibilities on product manufacturers to secure design.

It is not just governments who are concerned. A study indicated that senior managers are getting more worried about cyber security, with 82% rating it as a priority.[6] A further report suggested that 45% of people

DOI: 10.1201/9781003427681-2

believed their email or social media would be hacked within the year.[7] This compared with only 31% who thought they would need a trip to a hospital emergency department, so cyber security was more of a fear for them in their personal lives.

With respect to the obvious financial implications, it is also important to consider the indirect costs of cyber security incidents, like reputational damage. If customers lose trust in an organisation's ability to protect their data, they are likely to find an alternative supplier. One survey found that 87% of clients would withdraw their custom if they lost this trust.[7] Employee trust could also be affected, especially if they lose personal data as a result of the incident. Their personal contact details, payroll information, and passwords may have been compromised, putting them at risk of further attacks.

Another indirect cost is lack of access to work systems, as this could impact an organisation's ability to sell products and services. A 2020 report found the inability to conduct work because of a cyber-attack cost $590,000 and eighteen hours per department, on average, with only 32% of the cost recovered in insurance pay-outs.[8] Therefore, insurance fees also need to be factored into the overall cost of cybercrime. The same report identified an average incident response time of 19 hours. This figure should be multiplied by the number of personnel dealing with the incident, to achieve a truer cost. The UK Cyber Breaches Survey[6] found where money or data is lost, 38% of attacks took more than a day to resolve, so time spent responding to an incident also needs to be taken into consideration.

Commercially sensitive product information is also at risk. Cyber espionage is a form of cybercrime that seeks to obtain intellectual property for competitive advantage. This could result in cheaper imitation products on the market and an associated reduction in sales for the victim organisation.

Like commercial organisations, national security and our Critical National Infrastructure (CNI) are continuously under threat from cyber-attacks. In 2022, the National Cyber Security Centre (NCSC) (formed in the UK in 2016) advised organisations to be on "heightened alert" of cyber security attacks because of Russia's invasion of Ukraine.

Changes in CNI technology will affect cyber security. CNI organisations have traditionally been protected by 'air-gapped' systems, which have no connection to the internet. However, the US cyber security strategy recognises developments in green energy and the need to secure the design of associated tools, such as smart energy storage and cloud-based grid management. Advances in Operational Technology (OT), which includes industrial control and Supervisory Control and Data Acquisition (SCADA) systems, put this area at greater risk.

To conclude, the impact of cyber security incidents is significant. We will now consider what causes these incidents.

ATTACK VECTORS

Cyber criminals, or malicious actors, deploy various routes to obtain unauthorised access to computer systems and networks. Consistently over the years, malware (malicious software) has been the leading cause of cyber security attacks.[9,10] One type of malware is ransomware, where attackers demand money, often in the form of digital cryptocurrency, in return for restoring access to a victim's software and systems that they have blocked. The US strategy describes ransomware as "a threat to national security, public safety, and economic prosperity".[5]

The majority of malware is deployed by phishing attacks.[6,9,11] This can take the form of a link or attachment to an email. When the recipient opens the link or attachment, malicious software is installed and/or the recipient is tricked into supplying personal data, login information, or money. The data can be used to access an organisation's systems or other digital media associated with the recipient. 'Spear-phishing' emails (Business Email Compromise (BEC)) target individuals or organisations and are designed to look like they are sent from a trustworthy source. They are more believable because they often highlight information likely to be relevant to the user. They employ pretexting, where the attacker invents a story, or pretext, to enhance the credibility of the email. Similarly, 'whaling' emails are targeted at senior, high-profile individuals. Phishing emails employ tactics to encourage the recipient to activate the link or attachment, such as creating a sense of urgency, to respond before a negative consequence occurs or a positive consequence expires. Evidence suggests they are successful. Marret and Wright[12] found users respond to personalised messages that require an urgent response. Approximately one-third of recipients of a phishing email have been found to activate the malicious links.[17] However, people may be more likely to report phishing emails, with only 2.9% clicking them.[13] Technological advances may exacerbate the problem. Generative AI tools like *Worm GPT* can help criminals create mass phishing email attacks.[14] A study by Lim et al. reported that emails written by artificial intelligence (AI) were more successful than those written by humans.[15] Verizon's report[16] found that BEC attacks almost doubled in that year.

Other means of phishing include 'smishing', where the malicious link or attachment is delivered by text message, 'vishing'; voice phishing, which is conducted by phone and 'quishing', fraudulent Quick Response (QR) codes linked to malicious websites or content. AI and deep fakes (where the appearance and/or voice of the sender is altered to look like a trustworthy contact) have the potential to generate spear-phishing emails on a large scale.[17]

'Cryptojacking' is an attack route that uses phones or computers, hijacked via phishing emails, to mine for cryptocurrency. This allows the attackers to trade cryptocurrency using the victim's electricity.

Besides phishing, misconfigured servers and networks, including cloud databases, are another popular route to attack. IBM attributed most of the data compromises in 2019 to this,[11] and a survey predicted an increase in the impact of cloud-based attacks in 2023, compared to the previous year.[10] A reported shortage of skilled security personnel[5,10] may exacerbate this vulnerability. Failures in software patching, anti-virus protection, and use of default or poor passwords could also contribute to network incidents.

An increasingly common attack vector is via an organisation's software supply chain.[18] The EU 2023 Network and Information Security (NIS2) Directive[19] requires organisations to address their supply chain risks. However, the 2022 UK Cyber Breaches Survey[6] suggested that organisations failed to appreciate the importance of cyber security in supplier procurement.

Operational Technology (OT) in Critical National Infrastructure organisations has been highlighted as an area of vulnerability[5,9,20]. Although these control systems are traditionally 'air-gapped', they can be penetrated by USB-based malware, an attack vector that seems to be increasing in prevalence.[21] A USB device is designed to be used to transfer information between computers and peripheral devices. USB attacks rely on physical access to insert the device into equipment like OT. Therefore, it is important to consider physical site security as part of cyber security.

USB attacks typically occur via two main routes. Firstly, personnel who are authorised to be on a worksite can insert USB devices without knowing they contain malware. This may happen because they were given a USB device from a seemingly credible source, such as a conference organiser or supplier, without realising that the device has been compromised. Alternatively, they could find a device outside the secure area and be intrigued by its contents or see it as a useful means of transferring information for legitimate purposes. Secondly, malicious, unauthorised personnel can gain entry to a worksite and insert the USB device into computer systems. They can achieve this by 'tailgating;' following an authorised person through pass-managed security entry points. Sometimes attackers may use walking crutches or carry two cups so that they have their hands full. This encourages others to hold the door open for them out of politeness. In 2020, only 13% of OT attacks were found to be caused by insiders.[22] These insider attacks were apparently more reliant on malicious behaviours, such as attempting to sell sensitive information. The potential for outsider attacks needs to be captured and mitigated.

It can be hard to find written examples of OT incidents, and arguably, the industry needs to share incidents more openly. However, one known example occurred in 2010, when a malicious computer worm called 'Stuxnet' was thought to be behind a spate of attacks against Iran's nuclear sector.[23] These 'worms' replicate themselves and spread to other computers, but Stuxnet first relied on a person to install it into the OT system using a USB device. The infection caused the nuclear centrifuges to spin excessively, which resulted in deterioration.

The 'Internet of Things' (IoT), where different objects (such as household devices, fitness watches and, in a professional setting, IT and potentially OT) are connected to each other via the internet, presents another potential attack route. There is a desire to enhance the security of affected products.[5] IoT "listening" devices may also impact home-working security. 'Hybrid working', where people can work at home and in other locations as well as the office, increased during, and after, the coronavirus pandemic in 2020 and introduced new cyber vulnerabilities.[17] Information shared over video/tele-conferencing calls can be heard by others in the vicinity, including children and partners in the home (and their friends and colleagues if they are also on a call), and competitors or adversaries in other remote spaces such as cafes or trains. Electronic devices, such as laptops and wi-fi connections, that are inadequately protected, may be vulnerable to compromise. Paper-based information is also vulnerable if it is not adequately secured in lockable storage. It is also at risk of being lost in transit.

A 2022 UK Government report found that the use of personal devices (such as smartphones) by employees is common.[6] This means the organisation cannot control how well the devices are protected in terms of passwords and security software updates. It introduces vulnerability for the organisation and provides another attack route. Pirate movies and games downloaded on personal equipment can also contain malware.

The motives for these attacks are discussed in Chapter 3. For each attack vector discussed, there is a human element. In the next section, we discuss how the discipline of human factors can help mitigate the risk of cyber-attacks.

HOW CAN HUMAN FACTORS HELP?

Human factors (a term often used interchangeably with 'ergonomics') is a scientific discipline with roots in Psychology, Physiology, and engineering. It uses knowledge of human mental and physical strengths and limitations to design equipment, systems, or organisations around people. The aim is to enhance efficiency, safety, and the user experience. Human reliability is arguably more difficult to measure than machine reliability. The problem of how to predict human behaviours and vulnerabilities in cyber security is not easily solved by traditional cyber security methodologies. However, human factors may hold the key to this challenge.

Since the 1940s, just after World War II, human factors professionals have been addressing the issues that arise when equipment does not consider the people who use it. It was recognised that problems can arise when experts design for users who have a different experience or level of competence to themselves. Software developers, for example, are likely to possess a greater level of computer literacy than the majority of their users. End-users may operate equipment in ways the designers did not predict, and different people may use the same equipment in different ways. For this reason, it is essential to involve end-users in design.

In the field of cyber security, this means designing security procedures and equipment around the relevant people. If procedures are too strict, users will find workarounds. This is simply human nature and something that should be accepted. It is not usually malicious; the users are trying to find the easiest way to perform their job. For example, employees may often need to share legitimate information with external third parties. If cyber security procedures prevent or significantly inhibit, this information transfer, they may be tempted to use their personal email accounts or other potentially unsafe methods such as unauthorised USB devices. Therefore, failure to consult users about the security procedures that affect them creates vulnerability.

Human reliability analysis (HRA) is an area of human factors integration that considers human vulnerabilities in detail. It has been applied extensively in the safety sector, to support system safety cases and mitigate errors before they occur. Knowledge and lessons from this sector may be applied to enhance cyber security. HRA uses tools that can measure and predict so-called 'human error' in a given task and scenario. One such tool is the Human Error Assessment and Reduction Technique (HEART).[24] HEART incorporates Error Producing Conditions (EPC) that enhance the likelihood of human error. These include fatigue, competence, time pressure, task frequency, and familiarity. In the context of cyber security, people may read emails quickly because of the large quantity they receive, and time pressure to complete other activities. Their attentional resources may be diverted, increasing the likelihood that they will inadvertently click on a malicious link or attachment in a phishing email. Human factors and safety specialists know that apparent 'human error', such as this, cannot be considered in isolation. Incident investigation experience tells us there is a need to consider the chain of causation and potential organisational failings that could have contributed to the likelihood of a successful phishing email attack. For example, if the organisation's communications department or management team regularly sends emails that contain legitimate links or attachments, it might be difficult for employees to recognise a malicious email. If recipients are frequently under time pressure when they are processing emails, the risk is increased further. Spear-phishing or whaling emails create even more risk as they can be particularly hard to identify and can manipulate context bias, where judgement is affected by situational, background factors (see Chapter 4). If the recipient realises they have opened and activated the malicious attachment, they may be deterred from reporting it to their security team because of fear of punishment. Alternatively, they may be discouraged by an onerous reporting process, or they may simply not know how to report the incident or recognise the importance of doing so.

Training is often recommended to address assumed human failings. However, it can be patronising to mandate job training for an experienced employee who happens to have been involved in an incident. Training does not eradicate human error. Other factors may have contributed, such as time pressure or unsuitable procedures. While it is important to provide employees with training to enhance their awareness of cyber security threats, it is not the only solution. An understanding of human factors considerations can facilitate a more comprehensive assessment of cyber security risk and identify solutions to mitigate it. Post incident, human factors knowledge can help determine root causes and associated lessons to prevent recurrence. Detailed human factors solutions are described in Chapter 6.

Kraemer and Carayon,[25] found most security personnel blamed organisational factors for human error. Despite the need to consider systemic organisational factors and the chain of causation for cyber security incidents, "human error" can still get the blame. Some reports attribute as many as 95% of breaches to human error.[26,27] In 2023, Verizon[16] reported that the majority (74%) of all breaches included the human element and Cyber Magazine placed the figure at 82%.[28]

Terms such as "insider threat" suggest insiders, people within an organisation, are to blame for incidents. A 2016 IBM report[29] claimed that "60% of all attackers are insiders". However, it is advisable to analyse the root causes of incidents. A more recent, 2022, IBM report[30] attributed "human error" to just 21% of cyber security incidents, citing compromised credentials as an alternative cause. Another report in the same year claimed 82% of breaches had a human element.[13] In this case, the 'human element' included phishing vulnerability, misuse, and stolen credentials, as well as error. In any case, it is clear that addressing human vulnerabilities in an organisation, whatever the root cause, is essential to understand and mitigate cyber security risks.

Humans have been called the "weakest link".[31] However, they can also be a strength. Human flexibility and situational awareness provide strong defences against malicious actors. People may be able to detect suspicious behavioural indicators and patterns that technology cannot. We need to change the lexicon in order to properly address cyber security risks.

ATTACK EXAMPLES

In the 'Attack vectors' section, we learned about the Stuxnet cyber security attack on Iran's nuclear programme. This section will explore other cyber security incidents and highlight where human factors issues may have contributed. It can be hard to source detailed cyber security incident information, so the analysis relies on open-source material.

The first incident analysed concerns two banking organisations. In 2013, cybercriminals walked into branches of two high street banks.[32,33] They were able to convince staff that they were Information Technology (IT) engineers. This allowed them access to the computer system in each bank. From there, they installed a Keyboard, Video, Mouse (KVM) switch, a piece of equipment that facilitated remote access from another computer. From the remote computer, they accessed bank customer's personal financial information. £1.25 billion was reported stolen from the accounts of one of the banks. The criminals were later caught and most of the money was recovered. However, the banks likely suffered reputational damage and a reduction in customer trust as a result of the incident. Arguably, the main reason for the success of the attacks was the behaviour of the banking staff, in allowing criminals to access the banks' computers. However, it is interesting to note that the attack happened in two different banking organisations and branches. Therefore, it is hard to blame the individual staff members involved. It suggests that human factors considerations were at play. Two main psychological theories may explain the staff members' actions. Diffusion of responsibility[34] theory considers how people are influenced by the actions of others. They are less likely to help, or take responsibility for the situation, because they think someone else will do something. Because everyone present thinks the same thing, no one acts. This phenomenon suggests each bank staff member was deterred from challenging the fake IT engineer because no one else was challenging them. People also tend to trust people they like,[35] so if the attacker was amiable, staff may have been more inclined to believe they were genuine. Politeness and social compliance could have prevented bank employees from requesting proof of the attackers' credentials. Knowing these human vulnerabilities enables us to generate mitigating strategies. In this case, visitor management policy, where genuine IT engineers would be pre-booked and wear unique, visible identification, would have likely helped. Training staff about the incident, consequences, and other attack routes, would also be advisable. This incident illustrates that it doesn't matter how good technical controls are, it is essential to consider the human element in cyber security.

Media companies have also been targeted. In July 2020, the social media platform, Twitter (later called 'X'), was compromised.[36] Like the banking incident, the attackers apparently posed as IT engineers, this time using 'vishing'. They manipulated the Twitter employees' context bias by claiming to be calling about virtual private network (VPN) problems. This was credible because some employees had apparently been experiencing VPN problems since the move to remote working during the coronavirus pandemic. Several employees were persuaded to enter login details on a fake Twitter website. Some, but not all, employees reported the incident to their security team. The attackers were able to use the employees' credentials to take over the Twitter accounts of high-profile individuals including past and future US Presidents and celebrities. They posted fake messages from those accounts, claiming to offer to double bitcoin cryptocurrency donations made by their followers. The followers were directed to a fake bitcoin address, from which the attackers reportedly stole over $118,000. Cryptocurrency companies apparently blocked other attempted transfers to the value of approximately $1.5 million. In addition to the manipulation of employees' context bias, another human factor consideration is the organisational culture. There seems to have been a lack of senior management of cyber security at the time of the attack, as evidenced by the lack of Chief Information Officer (CIO) for seven months prior to it. A CIO was hired later in the year, after the incident. Management endorsement of cyber security helps create a positive culture, where employees are more cognisant of cyber security threats.

The communications company, Yahoo, suffered several cyber security attacks.[37] In 2013, reportedly one billion user accounts were compromised. The breach was not publicly announced until 2016, after other attacks on the organisation, including one in 2014 which used a spear-phishing email attachment. Attackers were apparently able to access customer's sensitive personal information including email addresses, passwords, and birthdates. The company was in the process of a sale to Verizon when the breaches were announced in late 2016. The sale price dropped by $350 million shortly afterwards.[38] According to a report,[39] the security team had previously warned the senior management of vulnerabilities, but they were internally nicknamed "The Paranoids" and their requests for funding had been refused. Managers were apparently worried about the impact on usability of additional security protection. This illustrates a need for a balance between usability and cyber security. A human factors approach to design would always consider usability and efficiency. However, where cyber security is needed, human factors and security professionals should work together, in conjunction with user representatives, to maximise usability while maintaining the required level of security. Cultural factors were at play too. Failure of senior management personnel to listen to the cyber security team encourages an attitude that can permeate throughout an organisation. It creates a shared belief that cyber security is unimportant. A human factors culture assessment can measure the maturity of a cyber security culture and identify solutions to enhance it.

A phishing email attack targeted accommodation in connection with the 2023 Eurovision Song Contest in Liverpool, UK. The travel company *Booking.com* apparently warned customers, after partner hotels had been compromised by inadvertently activating a malicious link in phishing emails. Attackers contacted the hotels' customers using the WhatsApp messaging platform and asked them to confirm their personal information, to avoid cancellation of the reservation. One reported victim was saved when their bank intervened over a fraudulent financial transaction.[40] It is therefore likely that the attack was motivated by money. The success of the attack was dependent on the human action of inadvertently activating a malicious link in a phishing email.

In October 2023, the British Library in London was submitted to a ransomware attack. Personal data, such as passport scans, of library customers and staff were stolen and (after the ransom was not paid), sold on the dark web. Damage to the library's infrastructure delayed the restoration of digital material, significantly impacting research for two months and continuing to affect it to a lesser extent after that. The attack was attributed to a phishing email or brute force attack on an IT supplier[41] (although server damage by the attackers apparently made the exact cause difficult to determine). This highlights the reliance on humans to identify phishing emails and the impact of weak passwords and supply chain vulnerabilities.

Ransomware also affected several London, UK National Health Service (NHS) hospitals in June 2024. Healthcare services were severely disrupted. Over 800 operations and 700 outpatient appointments had to be rescheduled.[42] A pathology laboratory supplier of the NHS was attacked. That meant blood tests results were affected, and test results (personal data) were stolen.

Weak passwords have been blamed for cyber security breaches. In 2017, a small proportion of email accounts in the UK Parliament were said to have been accessed in an attack by exploiting this vulnerability.[43] In 2021, the UK NCSC and US Department of Homeland Security (DHS) Cyber security and Infrastructure Security Agency (CISA) issued a warning, about weak passwords, to healthcare organisations associated with the coronavirus response.[44] The notice stated Advanced Persistent Threat (APT) groups (typically state actors) were targeting healthcare, coronavirus research, and pharmaceutical organisations, using the password spraying technique. Password spraying is a 'brute force attack', where attackers use the same commonly used password(s) to attempt to access multiple accounts. This trial-and-error approach allows them to access the accounts that use those common password(s).

Members of Parliament in the UK were again targeted, this time by a 'honeytrap' attack, in 2024. The victims were reportedly sent flirtatious messages and explicit photographs using a social media platform and encouraged to send some of their own. The attackers then threatened to share the compromising images and information unless privileged information was supplied to them. One of the victims reported the attack and the police investigated.[45] These types of attacks are sometimes called 'romance scams' and they prey on their victim's trusting nature and sexual and emotional needs and desires.

These are just a few examples of cyber security attacks that rely on human vulnerabilities. They illustrate that it is essential to consider human factors in cyber security. How exactly that can be achieved is addressed later in this book. Chapter 3 examines motivation and behaviours associated with cyber security.

KEY LEARNING POINTS

Key learning points from this chapter:

- Cybercrime is costly. It causes damage in terms of finance, reputation, loss of intellectual property, and damage to critical national infrastructure.
- Multiple attack vectors were described, all with a human element.
- The discipline of human factors uses established approaches to capture and mitigate human vulnerabilities and utilise human strengths.

Attack examples illustrated the need to consider human factors in cyber security.

NOTES

1 Petrosyan, A., 2024, Annual cost of cybercrime worldwide 2018–2029, *Statista*, https://www.statista.com/forecasts/1280009/cost-cybercrime-worldwide.
2 European Union, 2016, Regulation (EU) 2016/679 of the European Parliament and of the Council of 27 April 2016 on the protection of natural persons with regard to the processing of personal data and on the free movement of such data, and repealing Directive 95/46/EC (General Data Protection Regulation (GDPR)), *Official Journal of the European Union*.

3 ICO guide to data processing, penalties, https://ico.org.uk/for-organisations/guide-to-data-protection/guide-to-law-enforcement-processing/penalties/.

4 HM Government, 2022, National Cyber Security Strategy 2022, https://assets.publishing.service.gov.uk/media/61f0169de90e070375c230a8/government-cyber-security-strategy.pdf

5 The Whitehouse, Washington, March 2023, National Cyber security Strategy, https://www.whitehouse.gov/wp-content/uploads/2023/03/National-Cyber security-Strategy-2023.pdf.

6 HM Government, July 2022, Official Statistics, Cyber Breaches Survey 2022, https://www.gov.uk/government/statistics/cyber-security-breaches-survey-2022/cyber-security-breaches-survey-2022.

7 PWC, 2017, Consumer Series: Protect Me, https://www.pwc.com/us/en/advisory-services/publications/consumer-intelligence-series/protect-me/cis-protect-me-findings.pdf.

8 Smith, ZM et al. McAffee, 2020, The Hidden Costs of Cybercrime, https://companies.mybroadband.co.za/axiz/files/2021/02/eBook-Axiz-McAfee-hidden-costs-of-cybercrime.pdf

9 Burton, T, Thales, September 2022, Operational Security Trends to Expect in 2022 and Beyond, https://www.thalesgroup.com/en/united-kingdom/news/operational-technology-security-trends-expect-2022-and-beyond.

10 PWC, 2023, Cyber security Outlook 2023, https://www.pwc.co.uk/issues/cyber-security-services/insights/cyber-security-outlook-2023.html?WT.mc_id=CT1-PL52-DM2-TR3-LS4-ND30-TTA9-CN_Cyber security-DTI2023V3.

11 IBM X-Force Threat Intelligence Index 2020, https://www.ibm.com/downloads/cas/DEDOLR3W.

12 Marrett, K. and Wright, R., 2009, The effectiveness of deceptive tactics in phishing, *Proceedings of the Fifteenth Americas Conference on Information Systems*, San Francisco, California August 6th–9th.

13 Verizon, 2022, Data Breach Investigations Report (2008 – 2022).

14 Nachiappan, A., Sky News, September 2023, WormGPT: AI tool designed to help cybercriminals will let hackers develop attacks on large scale, experts warn, https://news.sky.com/story/wormgpt-ai-tool-designed-to-help-cybercriminals-will-let-hackers-develop-attacks-on-large-scale-experts-warn-12964220.

15 Lim, E., Tan, G., Kee Hock, T. and Lee, T., 2021, Turing in a Box: Applying Artificial Intelligence as a Service to Targeted Phishing and Defending against AI-generated Attacks, blackhat conference USA, 2021.

16 Verizon, 2023, 2023 Data Breach Investigations Report, OGWP1950623.

17 SoSafe Human Risk Review 2022, An analysis of the European cyberthreat landscape.

18 Gartner, 2022, 7 Top Trends in Cyber security for 2022, https://www.gartner.com/en/articles/7-top-trends-in-cyber security-for-2022.

19 EU, 2023, Directive on measures for a high common level of cyber security across the Union (*NIS2 Directive*).

20 Thales, 2022, Thales Data Threat Report, Navigating Data Security in an Era of Hybrid Work, Ransomware and Accelerated Cloud Transformation.

21 IBM, 2023, X-Force Threat Intelligence 2023, https://www.ibm.com/downloads/cas/DB4GL8YM.

22 IBM X-Force Threat Intelligence Index 2021, https://helpransomware.com/wp-content/uploads/2022/10/IBM-Security-X-Force-Threat-Intelligence-Index-HelpRansomware.pdf

23 Falliere, N., O Murchu, L and Chein, E, Symantec Security Response, November 2010, W32 Stuxnet Dossier, Version 1.3, https://www.wired.com/images_blogs/threatlevel/2010/11/w32_stuxnet_dossier.pdf.

24 Williams, J.C., 1992 'A user manual for the HEART human reliability assessment method', Nuclear Electric Plc., C2547-1.00.

25 Kraemer, S., and Carayon P., 2007, Human errors and violations in computer and information security: The viewpoint of network adminstrators and security specialists, *Applied Ergonomics* 38(2007), 143–154.

26 Cybint Solutions, 2020, Inside a hacker's mind - How to spot the weakest link in an organisation, (recorded webinar).

27 IBM X-Force Threat Intelligence Index 2015, https://www.ospi.es/export/sites/ospi/documents/IBM_2015_CyberSecurity_Intelligence_Index.pdf

28 Davies, V., 82% of al cyberattacks involve the human element, *Cyber Magazine*, April 12, 2023.

29 IBM X-Force Threat Intelligence Index 2016, https://www.catmanagers.org/product-page/ibm-x-force-threat-intelligence-report-2016

30 IBM Security, 2022, Cost of a Data Breach Report 2022, https://www.ibm.com/downloads/cas/3R8N1DZJ.

31 Accenture Security, 2019, The cost of cybercrime, Ninth annual cost of cybercrime study, unlocking the value of improved cyber security protection.

32 The Guardian, 13 September 2013, Cybergang foiled after allegedly hacking into London bank https://www.theguardian.com/uk-news/2013/sep/13/cybergang-hacked-santander-london-bank.

33 Sky News, 20 September 2013, Barclays Cyber Raid: Arrests over stolen £1.3m, https://news.sky.com/story/barclays-cyber-raid-arrests-over-stolen-1-3m-10433789.

34 Latené, B. and Nida, S., 1981, Ten years of research on group size and helping, *Psychological Bulletin*, 89(2), 308–324.

35 Eagly, A.H, Chaiken, S, 1984, Cognitive theories of persuasion. In L. Berkowitz (Ed.) *Advances in Experimental Social Psychology* (vol. 17). Orlando, FL: Academic Press.

36 New York State Department of Financial Services, Oct 2020, Twitter Investigation Report, https://www.dfs.ny.gov/Twitter_Report.

37 Trautman, L.J., and Ormerord, P.C., 2017, Corporate directors' and officers' cyber security standard of care: The Yahoo data breach, *American Law Review* 66, 1231–1291, https://papers.ssrn.com/sol3/papers.cfm?abstract_id=2883607.

38 New York Times, October 2017, All 3 billion Yahoo accounts were affected by 2013 attack, https://www.nytimes.com/2017/10/03/technology/yahoo-hack-3-billion-users.html.

39 New York Times, September 2016, Defending against attackers took a back seat at Yahoo, Insiders say, https://www.nytimes.com/2016/09/29/technology/yahoo-data-breach-hacking.html.

40 Rosney, D, 2023, Eurovision 2023. Hotel phishing scam targets song contest fans, https://www.bbc.co.uk/news/entertainment-arts-64822893.

41 British Library, 2024, Learning lessons from the cyber-attack. British library incident review, https://www.bl.uk/home/british-library-cyber-incident-review-8-march-2024.pdf.

42 NHS England, 2024, Update on cyber incident: clinical impact in South East London – Friday 14 June 2024, https://www.england.nhs.uk/london/2024/06/14/update-on-cyber-incident-clinical-impact-in-south-east-london-friday-14-june-2024/.

43 Reynolds, M., New Scientist, 2017, 'Cyberattack on UK Parliament exploited weak email passwords', https://www.newscientist.com/article/2138672-cyber-attack-on-uk-parliament-exploited-weak-email-passwords/.

44 National Cyber Security Centre, and Cyber security and Infrastructure Security Agency, 2020, Advisory: APT groups target healthcare and essential services.

45 Politico, April 2024, Naked photos senit in WhatsApp 'phishing' attacks on UK MPs and staff, https://www.politico.eu/article/uk-parliament-naked-photos-phishing-attacks-mps-staff/.

Motivation and behaviour

3

ATTACKER MOTIVATION

Attack vectors, or routes to cyber security breaches, were described in Chapter 2. Financially motivated cybercrime is common[6,13,16]. However, McAlaney et al.[46] included politics and the desire to create a dangerous self-image online, as attacker motivations. They described "impression management", where people aim to project a positive image online in order to be liked and appear capable. Other attacks may be carried out simply for enjoyment. The 'Lulzsec' hacker group, for example, was apparently motivated by fun; committing cybercrime "for the lulz". There is some evidence that malicious actors from outside an organisation can be influenced. For example, an experiment by Maqbool et al.[47] in 2016 found that providing financial rewards for hackers decreased the number of attacks. Nevertheless, organisations are likely to have more control over insiders (their employees) than malicious outsiders. Therefore, in addition to attacker motivation, it is also necessary to consider what drives the behaviours of *victims*, or employees with access to sensitive information, and malicious insiders. This chapter discusses theories of motivation, from the 1950s to the twenty-first century, and their relevance to cyber security.

MASLOW'S HIERARCHY OF NEED

Maslow's classic, 1954, motivation theory presents a hierarchy in a pyramidal structure.[48] Basic, physiological human needs, such as food and drink, are at the bottom of the hierarchy. They are followed, in the level above, by

DOI: 10.1201/9781003427681-3

the need for safety and security. Affiliation, or social, needs are next. This reflects the necessity of belonging to a group and having relationships with others. Esteem needs form the next, and penultimate, level in the hierarchy. This level includes self-esteem and recognition from others and is relevant to the work context in terms of praise and rewards for performance. The final layer, at the peak of the hierarchy, is self-actualisation, or a sense of fulfilment. Maslow theorised that the lowest needs in the hierarchy had to be satisfied before the higher needs were manifested, but this can be challenged. For example, lower needs, such as lunch, could be temporarily sacrificed or postponed in the pursuit of work and the associated recognition from others. Similarly, young adults could abandon the security of their childhood home in order to gain independence and self-esteem. However, the components of the model can provide useful insights into cyber security behaviour. For instance, the second level need for security suggests employees need to feel safe and protected from cyber-attacks. This could result in good behaviours, such as using anti-virus software and regularly installing software patches. It could also mean they become dependent on protection from their Information Security team.[49] Social needs, the next level, are important considerations in organisational cyber security. The extent to which good and bad cyber security-related behaviours are practiced in an organisation is affected by the social norm or culture. Organisational culture has been defined as a set of values to help members of the organisation know whether certain actions are considered acceptable or not.[50] Social norms are addressed later in this chapter.

Esteem, the next layer in the pyramid, could be adversely affected by falling victim to a successful cyber-attack. A victim may feel a loss of confidence in their own abilities and a perceived, or actual, loss of respect from their colleagues and managers. To avoid this, they may be deterred from reporting the incident. Consequently, their organisation may lack awareness of the vulnerability and fail to implement timely controls. In turn, this could result in further victims and data loss. Hence, organisations need to encourage a 'no blame' culture, whereby employees feel able to report incidents without fear of criticism or punishment.[51] Confidence in the ability to respond to cyber threats is discussed by Protection Motivation Theory (PMT) (which is discussed later). Ferris et al., 2009 (cited in Latham)[62] found that when self-esteem was associated with job performance, non-compliant behaviour was "highly unlikely". This emphasises the need to highlight the relationship between cyber security and job performance, for example in communications, rewards and training.

Cyber security does not seem to be an obvious driver towards self-actualisation or self-fulfilment, at the top of the pyramid. The associated *negative* consequences of a lack of cyber security may be more motivational. Similarly, physiological needs at the base of the pyramid appear to be less significant, although in extremes of poverty, staving off hunger is likely to be considered more important than the purchase of anti-virus software or the space to conduct sensitive work in private. Likewise, poverty may drive people to resort to cybercrime to achieve their basic human needs.

HERTZBERG'S DUAL-FACTOR THEORY

Hertzberg[52] in 1966, described a dual-factor theory of job satisfaction which comprises motivating and 'hygiene' factors. The latter are basic needs, a lack of which will cause a reduction in motivation, but, an increase, would not subsequently increase motivation at work. Examples of hygiene factors in the model are money, job security, status, policies, supervision, working relationships, and conditions. Motivating factors, on the other hand, include achievement, recognition, responsibility, advancement, growth, and the nature of the work itself.

Factors, such as job security, may be less relevant today, where changing employers is more common than in the past. Likewise, performance-related pay could incentivise job performance, which supports an argument that money can be motivating, and not purely a hygiene factor. However, the model may help explain why some employees might hold an expectation that they will be protected from cyber criminals as part of their working conditions, a hygiene factor, and therefore take cyber security for granted. It follows that if their cyber security is threatened, or they perceive that the organisation has a poor cyber security culture, their motivation could decrease. The model also suggests that recognition of good cyber security behaviours and/or performance would encourage these behaviours. Furthermore, it implies that when people are given more responsibilities over cyber security, for example by involvement in the design of procedures that affect their work, they are likely to feel more motivated.

INTRINSIC AND EXTRINSIC MOTIVATION

Intrinsic motivation theories highlight the importance of autonomy. Intrinsically motivated behaviours are performed for enjoyment, whereas extrinsically motivated behaviours are driven by external forces such as other people. For example, hacking for fun is intrinsically motivated, and performing a job you dislike is extrinsically motivated. Lai and Chen[53] found people who posted in online communities were intrinsically motivated to share knowledge, and 'lurkers', who observed but did not participate in the community, were extrinsically motivated by reciprocation. Vallerant and Ratelle[54] described four types of extrinsic motivation. In the first type, "*externally regulated*" behaviours are performed to gain a positive state or avoid a negative one. For example, cyber criminals can be motivated by money (a positive state), and employees can adopt good cyber security behaviours to avoid data loss (a negative state).

The second type of extrinsic motivation described is "*introjected regulation*", where the reasons for the behaviour are internalised but are still performed out of obligation. In "*identified regulation*", the third type, the individual identifies with the behaviour because they believe it to be valuable. Finally, "*integrated regulation*" means the behaviour is aligned with the person's other beliefs.

Amotivation, another type of extrinsic or non-self-determined motivation, is typified by a perceived lack of control over behaviours and the absence of a perceived link between behaviour and outcomes. Hence, an employee who believes they have no control over a cyber-attack and fails to see a link between their behaviour and cyber security may be less likely to perform good cyber security behaviours.

Three types of *intrinsic* motivation were identified by Carbonneau et al.,[55] namely, learning, accomplishment, and stimulation. They found the type of intrinsic motivation experienced partly depended on personality. People who were motivated by learning, engaged in an activity to gain understanding of a new area. Those who were motivated by accomplishment strove to master a task, and individuals who were motivated by stimulation sought a feeling of sensory pleasure.

Identifying factors that intrinsically motivate individuals allows us to shape cyber security training around them. The more cyber security behaviours are self-determined, the greater the likelihood of adoption. This means it is necessary to identify what individuals enjoy, intrinsic motivators. The theory suggests that people can follow security procedures out of a sense of duty or obligation, without fully appreciating the reasons for them. Therefore, training could seek to increase self-determination by educating employees about the *need* for the procedures, to enable them to believe in their value.

To illustrate this, imagine a scenario where a person enjoyed keeping in touch with their friends in other organisations by email. They occasionally discussed what they were working on, sometimes for social reasons and sometimes for work purposes. Security controls prevented the use of their work email account for communication with unknown external organisations, so they used their personal email account as a workaround. This practice continued for several years without incident, so the behaviour had become routine. When the dangers of discussing work on a personal email account, and associated need for the cyber security procedure, were discussed with them, they identified with, or even internalised, the need to protect work information shared by email. They stopped using their personal email account to discuss work. After consultation with the individual, the security team adjusted the procedure to allow communications with external organisations, where it was necessary for work purposes, using a controlled work email account. This enhanced organisational resilience by reducing the temptation for further workarounds. In this example, the original security control, preventing communication with other organisations, was external to the individual. They were not motivated to follow it until they understood the negative consequences that could occur. When they did, it allowed them to choose not to share work information using an uncontrolled email account. They felt more autonomy and were therefore more motivated to adhere to the recommendations of the security team.

Intrinsic motivation can vary by context.[54] A person might be motivated in one context (such as performing leisure activities) but not in another, such as their work environment. Work motivation in general can fluctuate. Li et al.[56] found employees showed lower motivation on Mondays. Interestingly, SoSafe[17] found that most phishing attempts were noticed by users between 7 am and 9 am on Mondays (and other mornings), so perhaps attackers assume people will be more vulnerable at this time, when their motivation is lowest. Alternatively, it may indicate that despite low work motivation, people may be more alert to phishing attempts in the early morning. Repeat exposure to activities that allow a high level of self-determination, or autonomy, has been found to affect contextual motivation.[54] For example, frequent experience of cyber-attacks could produce a sense of helplessness or amotivation. The challenge is therefore, to create positive associations with good cyber security behaviours.

SOCIAL INFLUENCES

Theory of Planned Behaviour

This model, proposed by Azjen in 1985,[57] purports that behavioural intentions are determined by attitudes, subjective social norms (see Maslow earlier), and the degree to which an individual feels they have control over the behaviour. Attitudes reflect an individual's viewpoint about a behaviour and the successful or unsuccessful performance of that behaviour. For example, a personal belief, or attitude, that cyber security is the responsibility of others and largely outside of one's control, may deter good cyber security behaviours. Social norms can be defined as underlying rules of behaviour in a society, and *subjective* norms are described as the belief as to whether other people approve or disapprove of the behaviour.[58] Both factors could influence the likelihood that an individual will perform a cybercrime or security-related behaviour. It explains why actions of hacker group members are influenced by the expectation of approval from others in the group. For organisations, this theory highlights the importance of the behaviour of peers and managers. If poor behaviours, such as sharing login information and propping open secure doors, are not criticised, the behaviours become socially acceptable. In other words, individuals may believe others approve, or at least do not disapprove, of the behaviours. Snyman and Kruger[59] found that people were more likely to share passwords if their colleagues did. Similarly, manager actions can influence culture and behavioural compliance.[60] If productivity is delayed because of adherence to cyber security procedures, and managers demonstrate disapproval of this, then the good cyber security behaviour is effectively discouraged.

The control aspect of the model could help explain why people may engage in workarounds to cyber security procedures. It suggests that people need to feel they can control the level of security, without it being imposed upon them in the form of overly restrictive procedures. For example, if security procedures and controls prevent an employee from sharing information legitimately with a third party, they may be tempted to find an alternative, less secure, route such as using a personal email account or unauthorised USB device. Therefore, it is important to involve representative users in the design of cyber security procedures.

The theory of planned behaviour (TPB) fails to consider the influence of variables such as emotions and unconscious bias on behaviour. Perhaps team members may understand the need to avoid discussing sensitive information in public but may become excited about their subject and talk about it anyway, so intent does not always predict behaviour. Likewise, someone may allocate time to read their emails, a behavioural intention, but become distracted by another task and, as a result, review their emails in a rush. Nevertheless, the concepts within the TPB, when used in conjunction with other models, may inform cyber security,[61] as discussed.

Social categorisation theory

Haslam 2004 (as cited in Latham,[62] proposed that employees are driven to maintain their sense of social identity in an organisation and categorise themselves in terms of either personal or group identity. Personal identity reflects the characteristics that an individual feels are important about themselves, such as gender, race, or profession. Group identity could include membership of a team, department, or union within a workplace, or a hacker group outside the workplace. Johnson[63] found people are more likely to open infected email attachments when they appear to be sent from someone they trust. Shared group identity may contribute to this.

According to the theory, group members aim to advance the social status of the group, but when employees see career advancement in other groups, their personal identity is prominent. This lack of group identity may help explain why malicious external parties are able to persuade individuals to betray their organisation in a social engineering attack. Conversely, when employees do not see a way out of a low-status group, they may reject senior management strategy. This has implications for compliance with cyber security policies and procedures. As people in these groups are often motivated by good working relationships (a Hertzberg hygiene factor),[62] this could be addressed by selecting cyber security champions from among their peers.

Competing values framework

Organisational culture types are explained in Cameron and Quinn's[64] framework. They identified "competing values" on two dimensions that influence organisational effectiveness. The first dimension is internal versus external focus. The second dimension is stability and control versus flexibility and freedom. From these dimensions, four organisational culture types were defined to support the Organisational Culture Assessment Instrument (OCAI). 'Adhocracy culture' reflects an external focus and high degree of flexibility and autonomy. These organisations value agility, creativity, and innovation. Conversely, 'Hierarchy culture' represents an internal focus and a large amount of stability and control. This culture advocates processes and procedures. 'Clan culture' reflects a flexible internal focus. This is a 'people-oriented' organisation, so social norms are likely to be important. A 'Market culture' has an external, customer focus with a lot of stability and control. Employees in this culture tend to be driven by results and productivity and are concerned about competition with the external market.

The four types of organisational culture affect cyber security behaviours in different ways. Adhocracies may be more resistant to strict, imposed cyber security processes and procedures. Hierarchical organisations would be more accepting of procedures, but employees may feel overwhelmed with the volume of procedures they already have. They may be less concerned with the threat of malicious external actors than externally focused organisations.

A clan culture is likely to be supportive and more resilient to factors that can cause malicious behaviours, such as a perceived lack of appreciation, or disagreement over management decisions. However, a high need for social compliance may create behavioural vulnerabilities such as sharing login information and passwords or holding doors open to allow unauthorised personnel access to secure areas. To enhance cyber security for this culture type, it is necessary to create social norms about best practice behaviour. Employees in a market culture need to communicate with, and monitor, external organisations as part of their business motivation. If not properly managed, this increases the risk of loss of information and susceptibility to phishing emails. This type of culture is more at risk of prioritising productivity over cyber security. The drive to protect internal competitive information is likely to help motivate individuals to practice good cyber security, however.

COM-B model

Michie et al.[65,66] devised the 'COM-B' model to understand and change behaviour. The model is comprised of three main components that determine behaviour: capability, motivation, and opportunity. It surmises that an individual needs to have the mental and physical knowledge and skills necessary to perform the behaviour (capability). They also need to have the intention, or motivation, to perform the particular behaviour over other behaviours. Finally, they need to be supported, and not constrained, by external factors in their physical and social environment (opportunity). The model states that motivation is influenced by capability and opportunity. The resultant behaviour then, in turn, influences capability, opportunity, and motivation, as illustrated in Figure 3.1.

Capability is further defined in terms of psychological and physical aspects. Psychological capability, in the context of cyber security, means the ability to detect phishing emails or remember passwords as well as knowledge of security risks. Physical capability barriers could include tiredness and physical stamina and their effects on mental concentration. Physical strength may influence the likelihood that an individual's laptop is lost because they become too tired to carry it and put it down.

The opportunity component of the model includes physical and social elements. Physical opportunity refers to environmental factors, such as time available to process emails or the provision of anti-virus software and other technical controls. The security of the physical environment is relevant here. Employees need private workspaces where they can discuss information that may not wish to share publicly. A procedure that prohibited the sharing of information with external organisations is an example of a physical opportunity barrier. Social opportunity is concerned with cultural norms.

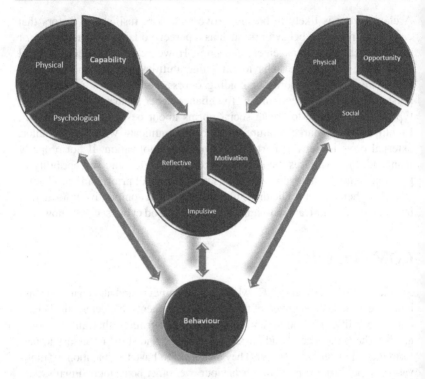

FIGURE 3.1   Representation of Michie et al.'s COM-B model.[65]

Motivation, in this case, incorporates decision-making, habits, and emotions. Reflective and automatic motivation are described. In the former, the individual analyses their relevant past experiences and plans their behaviour. This reflective, or slow thinking, could affect attitudes towards cyber security and perception of the value of compliance with security procedures. Automatic motivation, on the other hand, refers to more impulsive behaviours. Kahneman[67] described these processes as 'fast and slow' thinking. Automatic, or fast thinking, could help explain a lack of attention towards email management and the resultant likelihood of phishing email success, for example. It often applies cognitive biases and shortcuts. Targeted spear-phishing and whaling emails rely on contextual bias, where information relevant to the individual and their circumstances is used to increase the credibility of the malicious email. Decision-making and biases are discussed further in Chapter 4.

The model proposes that behaviour can be changed by altering at least one of the three main components. For example, if an individual's *capability* to perform good cyber security behaviours such as adopting a strong password, or recognising threats, is enhanced, then their *motivation* to perform the desired behaviours will increase. Similarly, if resources to support desirable cyber security behaviours, such as time for email management, anti-virus software, lockable storage, private workspaces, and a usable password policy, are provided, the *physical opportunity* to perform those behaviours increases. Changing an organisational culture so that cyber security is valued provides employees with the *social opportunity* to perform good cyber security behaviours.

Based on this model, if employees feel capable of performing good cyber security behaviours, and the organisational facilities and culture support this, their motivation to perform those behaviours is increased. The outcome of that behaviour performance then affects the capability, opportunity, and motivation towards it. For example, if the effort to perform the behaviour was greater than the reward or negative consequences, motivation would likely decrease. Likewise, if an individual employee does not observe a cyber-attack, they may lose the motivation to follow security procedures. Sharing knowledge about relatable cyber-attacks could help this.

FEAR MODELS

Health belief model

The Health Belief Model, by Hochbaum & Rosenstock in 1952 (cited in McKellar),[68] considers the motivating influence of fear and avoidance of negative consequences. The model postulates that good health-related behaviour depends on individual perceptions of threat susceptibility and severity. If the individual is then 'cued to action' by a trigger event, the likelihood that they will perform good behaviours is dependent on their perception of the benefits of the action minus the perceived barriers to, or level of control over, the action.

For enhanced cyber security, this would mean that individuals need to understand and believe threats and consequences. They need to trust that they can cope when they encounter a cyber-attack. Barriers to behaviour performance need to be removed, and the effort of performing good behaviours needs to be deemed worth it. Employees need to feel they can recognise a phishing email and/or report it quickly and easily. Onerous reporting mechanisms would be a barrier. This has implications for awareness training and incident reporting usability. Despite the omission of social influences, the model therefore provides useful considerations for cyber security.

Protection Motivation Theory

Rogers proposed the PMT in 1975. Like the health belief model, the theory assumes people are motivated by fear and the desire to protect themselves from negative consequences,[69] and a cost-benefit analysis is thought to take place. PMT states that people protect themselves based on threat and coping appraisal. Threat appraisal comprises an assessment of threat severity and likelihood, and perception of vulnerability to the threat. Coping appraisal considers the ability to successfully cope with the threat, and the cost of doing so. If threat severity, threat likelihood, and/or vulnerability are perceived to be low, people are less likely to employ protective behaviours, according to the theory. Similarly, if the perceived ability to cope with the threat, and/or the cost of doing so, is high, protective behaviours are less likely to be observed. Like reward theories (see later), perceived ability to comply with the behaviour, known as self-efficacy, was incorporated into the model. Bandura[70] discovered that an individual's perceived level of self-efficacy determined whether they would initiate and sustain coping behaviour. Self-efficacy can also determine behaviour performance.[71] There is evidence that training should seek to enhance self-efficacy and equip employees with the ability to cope with cyber security threats. Bishop et al.[72] found where individuals believe a cyber security threat is high, but feel they have the skills to protect themselves (self-efficacy), they exhibited safer cyber security behaviours. People with strong self-efficacy have also been found to report less frustration when faced with a computer crash/failure.[73]

Threats considered in the PMT theory were typically health-based, for example, eating unhealthily, drinking excessive alcohol, or lack of exercise; but they could also apply to cyber security behaviours. For example, cyber criminals who perceive the consequences of initiating an attack to be too great may be deterred. Conversely, they may attempt to create a sense of fear in phishing email content, to manipulate potential victims. The model indicates there is a need to ensure people are afraid of cyber security threats but feel equipped to deal with them in a cost-effective way. This needs to be factored into the development of security awareness programmes. Interpretation of the model suggests that the cost of performing good cyber security behaviours is likely to affect compliance, so onerous or restrictive procedures are to be avoided. There is evidence that social media and communication could be used to positively influence the adoption of good cyber security behaviours. Tang et al.[74] found that participation in Government social media influenced the uptake of protective security behaviours in relation to COVID-19 cons. They proposed that it affected individuals' perception of vulnerability and their belief that they could cope (self-efficacy) and successfully respond to the threat.

Using fear as a motivator needs to consider other factors, such as individual differences and personal experiences.[75] Marikyan et al.[69] found this helped explain why studies reported inconsistent findings in the application of the PMT model to cyber security. The adoption of protective behaviours can also be affected by the perception of moral obligation.[76] This suggests that if employees believed that the protection of the company's information is morally correct, they would be more likely to adopt good cyber security behaviours. This has implications for ethical policies and practices and employee engagement programmes.

Oakley et al.[77] expanded the PMT model to include ownership appraisal. In this context, it refers to the process by which an individual determines who is responsible for the protective actions. The study proposed that low adoption of domestic flood defences was associated with a belief that others were responsible for protective measures to prevent flooding. Human factors Cyber Vulnerability Investigations (CVI) found aligned results. Some employees believed their Information Technology/Security department would protect them from cyber-attacks. This belief allowed them to deflect the responsibility away from themselves and potentially reduced the likelihood that they would adopt protective cyber security behaviours. In an ideal organisational culture, all employees would take responsibility, or ownership, for cyber security.

Transactional model of stress and coping

A final fear motivation model is the Transactional Model of Stress and Coping.[78] It considers how people respond to potentially threatening situations and incorporates a primary appraisal stage, where individuals assess the level of the threat, followed by a secondary appraisal stage, where they analyse whether they have enough resources to be able to cope with the threat.[79] This suggests that if a cyber security threat is perceived to be dangerous, and a secondary appraisal concludes there are insufficient resources to cope with it, the individual will become stressed. Stress is known to inhibit performance and increase the risk of error,[24] so it might affect the response to a cyber security incident. This supports the earlier recommendation that training that equips individuals with not only the ability to cope with a cyber-attack but also the *belief* that they can cope with it, is likely to be beneficial.

The model postulates that stress can be overcome by changing either the threatening situation or managing emotions about the threat. The risk here is that individuals could emotionally 'downplay' the threat or accept it instead of dealing with it. However, they could alternatively adopt a more positive approach; choosing to change the threatening situation, by modifying their behaviours and adopting good cyber security practices. This has potential implications for training in cyber security threat management.

REWARD MODELS

Like fear models, reward models incorporate a cost-benefit analysis, but are driven by anticipation of positive consequences instead of negative ones.

Goal-setting theory

Locke, 1968, in McKenna,[48] considered the relationship between goals, motivation, and performance. They proposed that if people perceive a goal to be too difficult, the goal is unlikely to be motivating. Specific goals or targets, such as the number of people who respond to a phishing email, or lose a laptop, can influence organisational performance. Goal acceptance and commitment were added to the model later to further explain motivation and the effort made to achieve the goal. 'Acceptance' refers to how much an individual believes the goal is appropriate. 'Commitment' reflects the level of interest they have in the goal. Explaining the need for the goal was found to be effective (Wegge, 2000, in McKenna).[48]

Goal performance is affected by organisational support, individual abilities, and personality characteristics. Organisational support for cyber security goals could include management endorsement and provision of encrypted devices. Individual factors include the level of confidence in the ability to achieve the goal. If people feel a goal is an attainable challenge, they are more likely to work towards it than people who feel they are unable to achieve it. This is related to self-efficacy, described in fear models. Intrinsic rewards (e.g. a sense of achievement) and extrinsic rewards (e.g. promotion or recognition) result from the goal performance, and this affects satisfaction. Feedback was identified as enhancing creative effort towards goals.

Goal-setting theory has implications for cyber-security objective-setting as part of employee appraisal schemes. The model suggests that these objectives need to have specific targets, be achievable, relevant, and of interest to employees, supported by the organisation and be rewarded. For example, an objective about the number of lost laptops would be largely outside the control of non-security personnel. However, wearing appropriate company identification in the work environment would be achievable and within their power. The need for objectives should be explained to the individual to allow them to accept and commit to them. Successful performance can then be rewarded in the same way as other objectives.

Expectancy theory

Expectancy theory is based on the principle that people are motivated to perform behaviours if they perceive they will generate valued rewards. The amount of effort expended towards a behaviour is related to the probability of an associated reward (Vroom, 1964 in McKenna).[48] There is apparently a trade-off between the effort required, the likelihood of a reward and the value of that reward.[80] Although this seems to overlook the impact of emotion on behaviour, it helps explain why people accept website terms and conditions which allow the use of their personal data in exchange for a 'free' gift or financial reward. The effort required to read detailed terms and conditions may be perceived as greater than the immediate reward. Therefore, the less salient, potential risk of unwanted use of personal information, is outweighed by the effort required to read the necessary terms. Similarly, cyber criminals may be motivated by the anticipation of high gains for relatively little effort. They can manipulate the tendency for victims to continue following a course of action even when their suspicions are aroused, because of the amount of effort they have already expended. This tendency is known as the 'sunk-cost fallacy'.[81]

The model illustrates the need for good cyber security behaviours to be rewarded. The greater the perceived value of the reward, the greater the likelihood of these desirable behaviours. The value of the reward should reflect the amount of effort required to perform the good cyber security behaviours. Perceived probability of a reward is related to past experience. It is therefore important for managers to keep their promises and provide rewards when the required behaviour is observed. Rewards could be associated with performance against objectives (see goal-setting theory). To address the cost-benefits analysis process in the model, cyber security awareness training should use believable, high consequence, examples to illustrate that the cost of performing the good behaviours is less than the benefit of threat avoidance.

Equity theory

Equity theory (Adams 1963, in McKenna)[48] assumes that people need to feel they are fairly rewarded for their efforts compared to others. Perception of financial inequity may help explain why some people perpetrate cyber-attacks against those they perceive to be wealthier than themselves. Motivation to perform good cyber security behaviours is therefore dependent on the consistent and fair application of associated rewards between individuals and departments. If one department or individual is recognised for good cyber security behaviours and another is overlooked for similar achievements, this would give rise to a feeling of inequity in the unrewarded employees. This would likely cause a reduction in motivation and an associated deterioration of good security behaviours. It may even induce malicious insider activity.

SUMMARY OF MOTIVATION MODELS

Behaviour motivation influences described by the models in this chapter can be categorised into six main types, as illustrated in Figure 3.2: Theories of Motivation Summary (TOMS). Maslow's Hierarchy of Needs and Hertzberg's dual-factor theory proposed that behaviours are driven by specified human needs. The more control or autonomy individuals have over their behaviours, the more likely they are to perform them, according to the Dual-Factor Theory, Intrinsic and Extrinsic motivation theory, and the TPB. Fear and reward models reflect externally regulated, extrinsic motivation. These models also emphasised the importance of coping efficacy; the individual's belief that they can deal with the threat or achieve the goal. The Health Belief, PMT and Transactional Stress & Coping models attribute behaviour to fear of negative consequences or threat avoidance. Conversely, the Goal-setting, Expectancy and Equity theories suggest that behaviour is driven by rewards, or positive consequences, and effort. Social and subjective norms influence behaviour, as stated by the Hierarchy of Needs, TPB, Social Categorisation theory, Competing Values Framework and COM-B model.

As well as the influence of each of these factors on behaviour motivation, the outcome of the resultant behaviour can, in turn, affect the factors and motivation for future behaviour. Motivation is complex. Each behaviour shaping factor identified by the models could occur individually or in combination. Individual differences also need to be considered, as described in Chapter 4. A combination of the approaches is likely to yield optimal results.

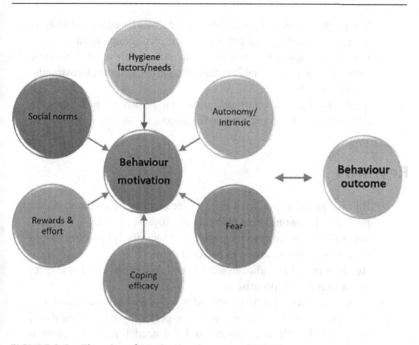

FIGURE 3.2   Theories of Motivation Summary (TOMS).

KEY LEARNING POINTS

Several findings and recommendations derived from analysis of motivation models have been categorised below and incorporated into Chapter 6.

Culture

- Cyber-attacks can damage self-esteem and deter victims from reporting attacks.
- Behaviour is influenced by the behaviour of managers and peers.
- Motivation can vary by context.
- Understanding of an organisation's culture type is necessary to identify tailored risk and mitigation strategies, as different types have different strengths and vulnerabilities.

- Social engineering, by malicious actors, works when individuals see career advancement outside of their own group/organisation.
- Barriers to desirable security behaviours, such as the physical environment, unreasonable technical controls, and time constraints, need to be removed.
- Social media and communication can positively influence the adoption of good cyber security behaviours.

Rewards

- Reward and recognition of good cyber security behaviours, and/or performance, would help motivate employees.
- The value of the reward for good cyber security behaviours should reflect the effort required to perform them.
- Rewards need to be allocated fairly and consistently and managers should keep their promises.
- Cyber security objectives should have achievable, relevant targets, be of interest to employees, supported by organisation and rewarded. People must understand the need for an objective to commit to it.

Incident management

- Companies should share knowledge about relevant cyber-attacks.

Policy & procedures

- Employees need to feel safe and protected from cyber-attacks. This could motivate them to adopt technical protections such as anti-virus software, encryption, and patching. However, it could also result in over-reliance on the IT department for cyber security.
- In extremes of poverty, cyber security may be considered less important than basic human needs, such as food.
- Allowing employees the autonomy to self-determine cyber security behaviours should increase compliance with those behaviours and reduce workarounds.
- The cost of performing good cyber security behaviours is likely to affect compliance.

Training

• Implications for training were discussed.

NOTES

46 McAlaney, J., Taylor, J. and Faily, S., 2015, *The Social Psychology of Cyber security*, Working papers of the sustainable society network+ Vol 3,1st International Conference on Cyber Security for Sustainable Society, February 26–27th, 2015, Conference Proceedings, Sustainable Society Network, ISSN 2052-8604.

47 Maqbool, Z., Makhijani, N. and Dutt, V., 2016, Effects of motivation: *Rewarding hackers for undetected attacks cause analyst to perform poorly*, Journal of Human Factors and Ergonomics Society 59(3), 420–431. https://doi.org/10.1177/0018720816681888.

48 McKenna, E., 2006, *Business psychology and organisational behaviour* (4th ed.). Psychology Press. ISBN 10: 1-84169-392-8.

49 Benson et al, 2018, cited in Ertran, A., Crossland G., Heath, C. and Jensen, R.J., 2018, *Everyday cyber security in organisations, literature review*, London: Royal Holloway University.

50 Moorhead, G. and Griffin, R.W., 2004, *Organisational behaviour: Managing people and organisations* (7th ed.). Boston, MA: Houghton Mifflin.

51 Dekker, S., 2016, *Just culture restoring trust and accountability in your organization* (3rd ed.). Boca Raton: CRC Press.

52 Hertzberg, F., 1966, *Work and the nature of man*. London: Staples Press.

53 Lai, H.-M. and Chen, T.T., 2014, Knowledge sharing in interest online communities: A comparison of posters and lurkers, *Computers in Human Behavior* 35, 295–306, ISSN 0747-5632.

54 Vallerand, R.J. and Ratelle, C.F., 2002, Intrinsic and extrinsic motivation: A hierarchical model, In E.L. Deci, & M.R. Ryan, (Eds.), *Handbook of Self-determination research* The University of Rochester Press , 2002, ISBN 1-58046-156-5.

55 Carbonneau, N., Vallerand, R.J. and Lafrenière, M.-A. K., 2011, Toward a tripartite model of intrinsic motivation, *Journal of Personality* 80(5), 1147–1178.

56 Li, R., Liu, H., Chen, Z., Wang, Y., 2023, Dynamic and cyclic relationships between employees' intrinsic and extrinsic motivation: Evidence from dynamic multilevel modelling analysis, *Journal of Vocational Behaviour* 140, 103813.

57 Ajzen, I., 1985, From intentions to actions: A theory of planned behavior. In J. Kuhi & J. Beckmann (Eds.), *Action-control: From cognition to behavior* (pp. 11–39). Heidelberg: Springer.

58 LaMorte, W.W., 2016, *Behaviour change models*, Boston: Boston University School of Public Health. https://sphweb.bumc.bu.edu/otlt/mph-modules/sb/behavioralchangetheories/BehavioralChangeTheories3.html.

59   Snyman D. and Kruger, H., 2021, *Group Behaviour in Cyber Security*, School of Computer Science and Information Systems, Potchefstroom, South Africa: North-West University.

60   Bruhn, A.B., Lindhal, C., Andersson, I. and Rosén, G., 2023, Motivational factors for occupational safety and health improvements: A mixed-method study within the Swedish equine sector, *Safety Science* 159, Falun, Sweden.

61   Sommestead, T. and Hallberg, J., 2015, The sufficiency of the theory of planned behaviour (TPB) for explaining Information Security policy compliance, *Information and Computer Security*, June 2015, 200–217.

62   Latham, G.P., 2012, *Work motivation history, theory, research, and practice* (2nd ed.). Los Angeles, London, New Delhi, Singapore, Washington DC: Sage.

63   Johnson. C.W. (2014) Anti-social networking: Crowdsourcing and the cyber defence of national critical infrastructures, *Ergonomics* 57(3), 419–433, https://doi.org/10.1080/00140139.2013.812749.

64   Cameron, K. and Quinn, R., 1999, in Bremer, M, 2012, *Organisational culture change, unleash your organization's potential in circles of 10* (1st ed.). Kikker Groep, ISBN: 978-90-819825-1-1.

65   Michie, S., Van Stralen, M.M. and West, R., 2011, The behaviour change wheel: A new method for characterising and designing behaviour change interventions, *Implementation Science*, April 2011, https://doi.org/10.1186/1748-5908-6-42.

66   Michie, S., Van Stralen, M.M., West, R., Social Change UK, 2019, A guide on the COM-B Model of Behaviour. https://social-change.co.uk/files/02.09.19_COM-B_and_changing_behaviour_.pdf.

67   Kahneman, D., 2012, *Thinking Fast and Slow* (1st Ed.). UK, USA, Canada, Ireland, India, New Zealand, South Africa: Penguin, ISBN-10 0141033576.

68   Mckellar, K. and Sillence, E., 2020, Chapter 2 in *Teenagers, sexual health information and the digital age*. New Castle: Elsevier Inc., https://doi.org/10.1016/B978-0-12-816969-8.00002-3.

69   Marikyan, D. and Papagiannidis, S., 2023, Protection motivation theory: A review. In S. Papagiannidis (Ed.), TheoryHub Book, ISBN: 9781739604400. http://open.ncl.ac.uk.

70   Bandura, A., 1977, Self-efficacy: Toward a unifying theory of behavioural change, *Psychological Review* 84(2), 191–215, https://doi.org/10.1037/0033-295X.84.2.191.

71   McAuley, E., 1985, Modelling and self-efficacy: A test of Bandura's model, *Journal of Sport and Exercise Psychology* 7(3), 283–295, ISSN: 0163-433X.

72   Bishop, L.M., Morgan, P.L., Asquith, P.M., Raywood-Burke, G., Wedgbury, A. and Jones, K., 2020, Examining human individual differences in cyber security and possible implications for human-machine interface design. In A. Moallem (Ed.) *HCI for cyber security, privacy and trust*. HCII 2020, Springer, Cham.

73   Bessiere, K., Ceaparu, I., Lazar, J., Robinson, J. and Shniederman, B., 2002, *Understanding computer user frustration: Measuring and modelling the disruption from poor designs*. Maryland: University of Maryland.

74   Tang, Z., Miller Z., Zhou, Z. and Warkentin, M., 2021, Does government social media promote users' information security behavior towards COVID-19 scams? Cultivation effects and protective motivations, *Government Information Quarterly* 38(2021), 101527.

75 Renaud, K. and Dupuis, M., 2019, Cyber security fear appeals: Unexpectedly complicated, *Proceedings of the New Security Paradigms Workshop*, pages 42–56, https://dl.acm.org/doi/10.1145/3368860.3368864.
76 Chen, M., 2020, Moral extension of the protection motivation theory model to predict climate change mitigation behavioral intentions in Taiwan. *Environmental Science and Pollution Research* 2712, 13714–13725.
77 Oakley, M., Mohun Himmelweit, S., Leinster, P. and Casado, M., 2020, Protection motivation theory: A proposed theoretical extension and moving beyond rationality—the case of flooding, *Water* 127, 1848.
78 Lazarus, R.S. and Folkman, S., 1984, *Stress, appraisal, and coping* (1st ed). Springer publishing. ISBN 0-8261-4191-9.
79 Goh, Y.W., Sawang, S. and Oei, T.P.S., 2010. The Revised Transactional Model (RTM) of occupational stress and coping: An improved process approach. *The Australian and New Zealand Journal of Organisational Psychology* 3, 13–20.
80 Bulgurcu et al., 2010, cited in Ertran, A., Crossland G., Heath, C. and Jensen, R.J., 2018, *Everyday cyber security in organisations, literature review*. London: Royal Holloway University.
81 Thaler, R., 1980, Toward a positive theory of consumer choice, *Journal of Economic Behaviour & Organisation* 1(1), 39–60.

Individual vulnerabilities

4

Chapter 3 identified that motivation theories cannot be considered in isolation. To understand and determine human behaviour, individual differences in people need to be considered. Personal attributes such as personality, attitudes, age, experience, and gender may affect their behaviour. Cognitive biases affect cyber security-related decision-making. Human error has been blamed for a significant proportion of cyber security incidents (see Chapter 2). However, human factors expertise tells us that this so-called "human error" is usually linked to wider, organisational factors,[25] such as those discussed in Chapter 5. For now, let us look at individual differences and associated cyber security vulnerabilities, starting with personality.

PERSONALITY

Some personality traits may be more vulnerable than others in terms of cyber security. Personality is commonly categorised by the 'big five' traits or the Five-Factor Model (FFM). The five traits are neuroticism, extraversion, openness to experience, agreeableness, and conscientiousness. The Revised NEO (from the original three domains: Neuroticism, Extraversion and Openness) Personality Inventory (NEO PI-R) by Costa and McCrae[82] captures and measures the five top-level traits (domains) and 30 sub-components (facets). After rating questionnaire statements using a five-point Likert scale, from 'strongly disagree' to 'strongly agree', individuals are scored to identify their position on the spectrum of each facet. Scores for any facet outside the middle of the range (where most people fall) are of most interest in the analysis of an individual's personality. Personality is said to be stable, but aspects can change over time, so a limit of 12–18 months is normally given for an assessment.

The neuroticism domain assesses the level of emotional stability. Calm, confident people score low in neuroticism; whereas anxious, angry, impulsive people score high. Individuals who score high in extraversion are sociable,

DOI: 10.1201/9781003427681-4

assertive, active, and loquacious. Carbonneau et al.[55] proposed that individuals who were motivated by stimulation (extraversion) sought a feeling of sensory pleasure. Introversion is at the opposite, low, end of the extraversion scale. People who score low in extraversion may be reserved and comfortable with their own company. They are typically independent, rather than crowd followers. Openness to experience reflects how curious and adventurous someone is about themselves and the external environment. Carbonneau et al.[55] confirmed that people who are motivated by learning (openness to ideas) engage in an activity to gain understanding. Low openness scorers tend to be conventional, traditional, and less emotional than high scorers. Agreeableness addresses the degree to which a person is sympathetic and supportive of others. High scorers tend to be trusting, trustworthy, and modest. Low scorers are more competitive, distrustful, and manipulative. People who score highly in conscientiousness are determined, reliable, and dutiful. Low scorers are less disciplined and lack organisation.

It is important to remember that people are complex and may score highly in one of the six facets within a domain, but low in another. Interpretation of personality tests also needs to consider the reliance on self-reporting and potential desire to answer according to the perceived expectations of the assessor. Too many or too few 'agree'/'strongly agree' responses should be treated with caution. Nevertheless, it is interesting to profile personalities in terms of cyber security vulnerabilities, using psychometric tools like the NEO PI-R.[83] Certainly, Shappie et al.[84] found a significant association between conscientiousness, agreeableness and openness, and self-reported cyber security behaviours. They recommended improving self-efficacy to enhance cyber security. This could be achieved by training. Similarly, McBride et al.[85] found conscientious individuals with a perception of low risk were less likely to violate cyber security policy. Conscientiousness has also been associated with good password behaviours.[86]

A high level of impulsiveness (a facet of neuroticism) could affect the likelihood of responding to malicious links or attachments in phishing emails. Indeed, Egelman and Peer[87] and Raywood-Burke et al.[86] inversely correlated good security behaviours with impulsivity. The latter study also found a small correlation between low neuroticism (which includes impulsiveness as a facet) and good cyber security behaviour, specifically, device securement. Conscientious deliberators may be less vulnerable in this respect. Other components of high neuroticism may be associated with *good* cyber security behaviour. Gross et al.,[88] found levels of anxiety in response to cyber threats were like those provoked by traditional terrorism acts, like bombings. A high level of anxiety might make someone more likely to identify a malicious phishing email. McBride et al.[85] found policy violation was more likely in high neuroticism scorers who thought punishment was unlikely, demonstrating that personality should not be considered in isolation from situational factors.

Extraversion may also create risks. McBride et al.[85] found that extraverts were more likely to violate cyber security policy. People scoring highly in 'warmth', an extraversion facet, might be more likely to (unwittingly) befriend cybercriminals. Extraverts may also be inclined to click unsafe links out of boredom. However, some facets of extraversion may be beneficial to cyber security. Independence, a characteristic of low extraversion, has also been found to be positively associated with cyber security.[87] It could be argued that high assertiveness, an extravert characteristic, may afford individuals more resilience to manipulative social engineering attacks. Analysis should therefore not purely focus on the *domains* of personality, but it should also address the relevant component facets.

Openness is another potential vulnerability. Someone highly open to values may be more likely to join a cybercriminal group aligned with a political cause. Alseadoon et al.[89] found people with high openness (and extraversion) were more likely to adversely respond to phishing emails. They may be more likely to reveal information online. High scorers in this facet are less likely to accept authority, and therefore, one assumes cyber security policy. On a positive note, Egelman and Peer[87] found inquisitive participants were more likely to report that they followed better security practices. Similarly, Raywood-Burke et al.[86] found a positive relationship between openness and specific good cyber security behaviours: "proactive awareness" (including checking links and websites), updating software, and good password management. Therefore, perhaps openness to ideas can extend to cyber security as a concept. The impact of situational factors also needs to be considered. McBride et al.[85] found open individuals were more likely to comply with policy when they had low self-efficacy, an anticipation of low risk, or a perception of few or no negative consequences associated with policy compliance.

Highly trusting (agreeable) people are apparently more vulnerable to social engineering attacks from malicious parties.[90] There were correlations between agreeableness and persuasion by authority, people they like, reciprocation (feeling the need to return a favour), and commitment (keeping promises). These persuasion tactics are employed by cybercriminals to manipulate potential victims.[91] Low straight straightforwardness, (another (low) agreeableness facet), is compatible with criminal social engineering manipulation and deception. Other aspects of agreeableness have been associated with positive cyber security behaviours. People scoring high in compliance (part of 'agreeableness'), (and those with a high sense of duty (a conscientiousness facet)), may be more likely to follow an organisation's cyber security policy and procedures. Indeed, McBride et al.[85] found agreeableness in general to be associated with good cyber security policy compliance when combined with low self-efficacy and low perceived punishment severity. Raywood-Burke et al.[86] correlated agreeableness with a specific aspect of good cyber security behaviour: proactive awareness. Agreeableness therefore has advantages and disadvantages in terms of cyber security, so focus on sub-facets of the domain is necessary.

Hogan, 2004 (in Latham)[62] proposed that individual differences can be explained by personality theory. Their taxonomy comprises three main human needs: acceptance and approval; status, power, and control; and predictability and order. Through these needs, the theory aims to explain people's ability to get along with others and get ahead: drivers of motivation and behaviour. The need to get along corresponds to the neuroticism, agreeableness, and conscientiousness personality domains, and the need to get ahead is associated with extraversion and openness to experience. The extent of a person's need for acceptance and approval might drive how well they follow social norms, including cyber security culture. The need for power and control echoes the finding, as discussed in Chapter 3, that the more control or autonomy individuals have over their behaviours, the more likely they are to perform them. Finally, the third need for predictability and order may help explain individual reactions to disruption caused by cyber-attacks. Hogan personality inventories[92] consider how people relate to others; their intrinsic motivation; and development or risk areas.

Example traits from the Hogan PIs that could benefit cyber security include high adjustment (tendency to remain calm under pressure); low interpersonal sensitivity – willingness to confront others; high scepticism and high cautiousness. Potentially detrimental traits include high sociability. People in this category are talkative and attention-seeking. Like the NEO PI-R extraverts, they may be more likely to share information and become targets for spear-phishing or whaling attacks. Low scorers in the prudence scale are spontaneous and open-minded. As described earlier in this section,[89] this may make them more vulnerable to phishing emails, especially those that create a sense of urgency and allow little time for a response. Those scoring low in power tend to avoid confrontation, so they may be less inclined to question a social engineering attacker. Individuals with a low 'tradition' score seem to prefer autonomy and are more likely to challenge security procedures. It is particularly important to consult people with this personality regarding the implementation of new procedures. The Hogan Motives, Values and Preferences Inventory (MVPI) also measures commerce, which includes the degree to which people are financially motivated. Given the large amount of cyber-attacks motivated by money,[6,13] this could feature in the profiles of some cybercriminals. Low diligence includes poor attention to detail, which may also be a factor in the likelihood of spotting a phishing email. People scoring low in the dutiful scale may resent authority and, as a result, be less responsive to imposed cyber security policies and procedures. A summary of mapped positive, desirable cyber security traits extracted from the NEO PI-R and Hogan PIs is presented in Table 4.1. For cybercriminal traits, read the opposite.

Arguably, another dimension of personality could be added to the table. Lee and Ashton[93] expanded the NEO with the addition of Honesty-Humility, (also known as the "H-factor"), to create the HEXACO Personality Inventory. Other HEXACO domains are similar to those in the NEO PI-R, so the acronym is derived from the components: Honesty-Humility, Emotionality (similar to the NEO's Neuroticism), eXtraversion, Agreeableness, Conscientiousness,

TABLE 4.1   Profile of desirable cyber security personality traits

NEO PI-R	HOGAN PI
High conscientiousness	High diligence (good attention to detail); high dutiful
High sense of duty (conscientiousness)	High prudence (dependable, self-controlled)
Low impulsivity (neuroticism)	High cautious – risk averse
High anxiety (neuroticism) about cyber threats	High adjustment (steady under pressure)
Low warmth (low extraversion)	
High independence (low extraversion)	
Low gregariousness (low extraversion)	Low sociability and low affiliation (prefer to work alone); low recognition (avoid the limelight)
Low excitement-seeking (low extraversion)	Low hedonism (low desire for fun and excitement)
High assertiveness (extraversion)	High power (willing to confront others)
Low trusting (agreeableness)	High scepticism; low interpersonal sensitivity (willingness to confront others)
High straightforwardness (agreeableness)	
High compliance (agreeableness)	Low mischievous (compliant); high tradition – low autonomy and high compliance with procedures
High altruism (agreeableness)	High altruism
	Low commerce (not financially motivated)

and Openness to experience. Facets of the new 'H' domain test sincerity/ manipulation, loyalty, fairness, ethics, interest in money & possessions, modesty, and sense of privilege. Some facets (such as modesty and straightforwardness) were previously captured in the NEO PI-R. Lee and Ashton argue that people who score high in Honesty-Humility do not typically exploit others, a potentially important factor in understanding cybercriminal personality.

The complexity of personality means the profile should only be used as a guide, and further validation is advised. Some traits could be either an asset or a vulnerability to cyber security, depending on the context. For example, openness was excluded from the table because it has been associated with both positive and negative behaviours. Inquisitiveness (an openness trait) could cause someone to open a malicious email, but it could also encourage them to investigate cyber security concepts, as described earlier. Furthermore, some of the positive traits in the table have also been observed in cybercriminals. Kranenbarg et al.[94] found that suspected cybercriminals were less extraverted, more conscientious, and more open to experience than other "offline" criminals. They were more like a non-criminal group in this respect, when tested using the HEXACO PI. Raymond, 2015 (in Patterson and Winston-Proctor),[95] likewise referred to hackers as introverts, interested in a wide range of intellectual topics. In Kranebarg et al.'s study, both criminal groups scored lower in honesty-humility than the non-criminal group, as one might expect, although the difference between cybercriminals and non-criminals was not statistically significant. The cybercriminals scored more similarly to the non-criminal sample in terms of high patience, perfectionism, and prudence; traits which the study authors argued facilitated cybercrime. Like other criminals, they scored low in modesty (a facet of honesty-humility in the HEXACO), fearfulness, and flexibility. Interestingly, the cybercriminals also scored significantly more highly in diligence than both other groups (offline criminals and non-criminals). This suggests attention to detail is beneficial to hacking and other cybercrimes.

ATTITUDES AND BEHAVIOUR

Closely aligned with personality and motivation is attitude. A person's attitude towards cyber security affects how they respond to cyber threats and interventions. Malicious insider behaviour has been said to be attitudinal. Dalal and Gorab in Zaccaro et al.[96] claimed some employees are predisposed to "Counterproductive Work Behaviour"; that is intentionally detrimental to the organisation.

Attitudes and behaviour are shaped by our experiences. Safa et al.[97] discovered that employee attitudes to cyber security procedure compliance were significantly affected by experience, cyber security knowledge sharing, collaboration, and intervention. Unsurprisingly, employee attitudes about cyber security policy were found to significantly affect their intention to comply with it. The effect of previous experience on attitudes and behaviour is further illustrated by Gelbrich et al.,[98] who showed that customers' reactions to a hotel service failure compensation varied according to the strength of their relationship with the hotel. Customers with a strong relationship reacted better to the compensation than those with a weaker one. This suggests that employees with a strong bond with their organisation might respond better to cyber security policy and procedures. Family attitudes have also been shown to impact behaviour. Van de Weijer and Moneva[99] found cybercriminals (and other criminals) were more likely to have criminal parents and siblings than non-criminals.

A casual attitude to online cyber security can be caused by the disinhibition and invisibility effects of the internet. People have been shown to disclose more information or act differently online than they would do in person.[100] When browsing websites, it can be tempting to accept pop-up messages without reading lengthy terms and conditions, or caveats. Morgan et al.[101] found that subjects spent very little time reviewing pop-up messages and had a high propensity to accept, rather than decline them. Training, specifically Malevolence Cue Identification Training (MCIT), was found to significantly reduce the number of malicious pop-ups accepted. Training does not always address individual differences, however. Vishwanath et al.[102] reported that individual email habits seemed to override phishing email training.

When people hold conflicting attitudes and behaviours, this is known as cognitive dissonance.[103] For example, an individual may recognise that cyber security is important, but continue to ignore associated procedures because they find them time-consuming. To resolve the cognitive dissonance, they may convince themselves that they are unlikely to fall victim to an attack. Training needs to explain the chances of a personal attack.

AGE AND EXPERIENCE

Adding to the idea that people's behaviour is shaped by their experiences, the impact of age on cyber security has been studied. Ben-Asher and Gonzalez[104] found that the more cyber security knowledge subjects possess, the greater their detection of malicious network events. The knowledge also apparently allowed them to better classify benign events. Similarly, Wright and Marett[105] identified that web experience helped determine whether participants provided sensitive information in response to a phishing email. Emotional benefits may potentially arise from experience too. Bessiere et al.[106] found that the more perceived experience a person had, the better their mood after encountering a computer problem.

Age effects were identified by Morgan et al.[107] Their study found that the majority of older participants (aged 60 years and over) accepted fake pop-up messages that were designed to mimic those from trusted organisations when performing a mentally demanding memory recall activity. The older adults were roughly ten percent more likely to accept than younger participants (with an average age of 18) in a previous iteration of the study. When the older adults were allowed unlimited time in which to rate pop-up messages, their results were more similar to those of the younger adults, suggesting time pressure was an issue for the older people in particular.

Personality can also vary with age. Costa and McCrae[82] found older people scored a little lower on neuroticism, extraversion, and openness than younger people, and a little higher, on agreeableness and conscientiousness. This suggests that older people would be more vulnerable to cybercrime in some respects, in that they are more likely to want to please and trust attackers (agreeableness). However, they would be less likely to respond impulsively to phishing emails, for example, and more likely to follow cyber security policy and procedures (conscientiousness). This mixture of vulnerabilities was further illustrated by Branley-Bell et al.[108] They found that older users were less likely to lock their devices, but more likely to have proactive risk awareness and perform good cyber security behaviours regarding passwords and installing updates. Therefore, research suggests that age cannot be reliably used as a broad indicator of cyber security risk.

GENDER

Research has found mixed results in relation to gender and cyber security. Henley et al.[109] found 94% of intellectual property breaches were committed by male employees. Likewise, Cappelli et al.[110] reported that internal IT sabotage attacks were male-dominated. However, this could be related to the relative proportions of men and women in the workplace, rather than a cyber security gender difference. The majority of the employees in Henley's review were engineers or scientists; roles, it could be argued, traditionally associated with a male majority. Indeed, studies by Branley-Bell et al.[108] and

Raywood-Burke et al.[86] found no significant relationship between gender and cyber security behaviour.

There may be gender implications for the design and success of smishing (malicious texting attacks). Kato and Kato[111] studied mobile phone text message response speed indicators. They found that women tended to refer to the content of text message, whereas men tended to refer to their own situation and attributes of the sender.

Gender is a factor in personality, according to Costa and McCrae.[82] They reported that women tend to score more highly in the neuroticism domain, in particular the anxiety sub-facet. This could make them more worried about cyber-attacks. Women were also reportedly more agreeable, scoring relatively highly in the straightforwardness and altruism facets. This would make them not only more sincere and trustworthy but also more likely to want to help people, including (unwittingly), cybercriminals in social engineering attacks. Hence, gender differences apparently have advantages and disadvantages in terms of cyber security.

DECISION-MAKING

Reflective and automatic decision-making were described as components of the COM-B model of motivation in Chapter 3, with reference to Kahneman's fast and slow thinking concept[67] and the implications for cyber security. Wickens'[112] decision-making model states that people develop a hypothesis based on cues from their environment. The hypothesis is derived from a combination of memory of past experiences and short-term, working, memory. Alternative hypotheses are evaluated in response to situational factors or cues. Outcomes of actions based on the final hypothesis, in turn, affect future hypotheses in similar scenarios. This decision-making process is further influenced by cognitive biases and perceived risk. It helps explain online behaviour and how people respond to cyber security threats, as well as the decision, or not, to adopt good cyber security behaviours. This section further explores bias, risk-taking, group decision-making, and the implications for cyber security.

Cognitive biases

Cognitive biases are essentially prejudices derived from heuristics: mental shortcuts. People do not need to remember and consider all possible information about everything they encounter. For example, when taking a familiar car journey or walk, we do not necessarily pay attention to sign-posts or subtle changes to landmarks because that information is not required to complete the goal of reaching the end destination. Heuristics are therefore a way of simplifying the vast quantities of information in our environment, and everyone is susceptible to the associated cognitive biases. Many heuristics and types of cognitive bias have been defined. Some, relevant to cyber security are described next.

The **anchoring** heuristic, as described by Tversky and Kahneman,[113] refers to an initial piece of information that is used as a starting point, or 'anchor', and then adjusted. Issues arise when the adjustment is insufficient and new information is not assigned the same level of importance as the initial anchor. The order in which information is received can therefore affect a decision. Anchoring can bring about **confirmation bias**. This means that any information received that contradicts the initial hypothesis is downplayed or dismissed in favour of information that supports it. An air accident involving First Air Flight 6560 in Canada, in 2011, was partially attributed to the pilot's confirmation bias.[114] The pilot, incorrectly believing the autopilot was working, disregarded information from indicators that correctly suggested the aircraft was off course on its approach to landing, even when the information was reiterated by the first officer, and ultimately crashed into terrain, killing most of the people on board. Life-threatening consequences of cognitive bias have also been observed in healthcare. Saposnik et al.[115] found anchoring (as well as information availability, over-confidence, and risk tolerance) was associated with errors in medical diagnosis. It stands to reason that confirmation bias could have consequences for cyber security too. For example, if an employee believed a malicious social engineering attacker, website or email was genuine, additional information to the contrary may be rejected.

Contextual bias can help explain the success of targeted phishing email campaigns. Decision-making is affected by contextual information.[116] Cybercriminals target key employees because of their access to sensitive information. If one of those employees had recently attended a conference, for example, cybercriminals could contact them through their email address on the delegates list or by social media platforms. They could enclose an attachment containing malware, perhaps under the guise of a paper they apparently thought the recipient might be interested in, given their interest in the conference subject area. This is a realistic scenario, based on the recipient's contextual environment. The recipient could open the attachment in good faith, thereby creating vulnerability for their organisation.

A business context may demand that emails are processed quickly. People tend to attend to information that is most obvious, particularly when under time pressure; a phenomenon known as **"salience bias"**.[117] This helps explain why less apparent, suspicious aspects of a phishing email could be missed, and has implications for the design of security applications and alert management. The most important information needs to be the most prominent in order to draw the user's attention.

Similarly, the **availability heuristic** describes how information at the forefront of our mind is assumed to be more likely or frequent.[118] Availability of information can be affected by the emotional reaction it evokes. A cyber security attack with significant consequences may more easily come to mind than one with no personal repercussions. Likewise, an incident with more time investment, such as loss of work data, may have more meaning and be more easily recalled. This suggests employees may be more likely to correctly identify a suspicious email or social engineering attack if they had recently received cyber security awareness training, or heard about an incident, and the warnings were at the forefront of their mind. Furthermore, the value of work data generated by an employee may be worth more to employees or organisations than to cybercriminals who try to steal it, according to principles of the "**endowment effect**".[119]

The design of applications and equipment is also affected by biases. Ting[120] described the effects of pro-innovation bias and the bandwagon effect on cyber security. The **pro-innovation bias** can cause organisations to develop technology without properly considering security in the design. The **bandwagon effect** assumes that as the number of people who adopt a new technology increases, so does confidence in that technology. Research and development of human factors in cyber security therefore needs to keep pace with technological developments to mitigate these biases. Likewise, new technologies need to consider the human element in cyber security by design.

Stereotypes are associated with the "**representative heuristic**".[118] The theory suggests that people make judgements based on how well informational cues are representative of a known cause. For example, in person, a cyber security attacker does not necessarily look like the perceived image of a criminal. In the banking attack described in Chapter 2, criminals successfully posed as IT professionals partly because of this. In the same way, suspicious computer behaviour, such as slow running and crashing, could be mistakenly blamed on poor equipment or connection speed, instead of a cyber-attack, because it is representative of both.

A lack of confidence about cyber security may cause someone to seek more information. Conversely, **overconfidence**, has been linked to safety accidents.[121] It could result in a blasé attitude towards cyber security and create neglectful or dangerous behaviours. Confidence could be increased by training, and overconfidence could be addressed by sharing relevant, realistic attack examples.

The **optimism bias** means people tend to underestimate the risk of negative occurrences happening to themselves compared to others. In other words, there is a tendency to think that bad things happen to other people. Alnifie and Kim[122] discovered that the optimism bias affected the perception of cyber security risks and associated risky behaviours, such as sharing passwords and failing to follow security procedures. Risk-taking behaviour is discussed more later.

Security language can also affect decision-making. Dramatic words such as "fatal" and "firewall" may desensitise users. A "brute force attack" for example, may sound aggressive, but actually have lower damage potential than an "infection" of malware. Choice of language has been shown to affect the success of warning messages. Carpenter et al.[123] found the word "hazard" in a warning message was more effective than "caution" and "warning" in reducing email address disclosure. Describing cyber-attack victims as having "fallen for a scam" attributes blame on the victim instead of the attacker.

Knowledge of heuristics and biases supports cyber security incident investigation. This wisdom helps determine the root causes of incidents and facilitate the generation of mitigating strategies. It is important to remember that everyone is susceptible to biases, so blaming individuals is of little value compared to consideration of the wider root causes.

Risk-taking

Another aspect of individual vulnerability of potential relevance to cyber security is risk-taking. Studies indicate that people choose loss aversion over gain.[124] When faced with a choice between a certain loss or an uncertain, but greater loss, people are likely to choose the second, more risky, 'uncertain' option. However, if the scenario is framed as a gain, the less risky option is typically selected. This tendency is particularly relevant to cyber incident response. If an organisation's executive team is faced with a potential loss of data because of a cyber-attack, they may have the choice of shutting down internet systems and productivity, or continuing and attempting to resolve the problem. The typical, loss aversion, decision-making process would drive them to continue, rather than shut down. The amount of *effort* invested in a course of action can also drive people to continue with it, even though it is inadvisable. This tendency to 'throw good money after bad' was termed the 'sunk-cost fallacy'.[81] Response training therefore needs to highlight these decision-making tendencies and evaluate their impact.

Judgement of risk is further complicated by the finding that people generally underestimate the probability of events, unless they are very rare.[112] Imgraben et al.[125] found most people questioned underestimated cyber security risks associated with mobile phones. People also downplay the cumulative effects of risk, so training and communications need to highlight the long-term risks of poor cyber security. Probabilistic judgement is also affected by the media, and internal communications, activating the availability bias. So, news of a recent cyber-attack will likely increase the perceived risk of further cyber-attacks.

Personal experience also has a part to play in risk perception. If someone has no experience of a cyber security incident, they will typically perceive it to be lower risk. Similarly, if they *have* experienced an incident but there were few, or no, adverse consequences, they may continue to adopt poor security behaviours. However, experience also has advantages. Experienced command team members utilise naturalistic decision-making strategies. They can adopt cues from their environment and utilise more hypotheses and action patterns based on their past experiences. Expert decision-makers have been found to correlate, or group, related information to save time where fast, high consequence decisions are required. Naturalistic decision-making is difficult to observe in a laboratory, so analysis largely relies on applied human factors engineering. Human factors design should also consider decision aids to support complex information processing. As well as making people aware of biases, incident response training should provide immediate feedback about the consequences of actions and emphasise correlations between information and actions.

Fast decision-making, as employed by expert decision-makers, can be measured by personality testing. The NEO PI-R described earlier, includes "deliberation" as a facet under conscientiousness. Someone scoring low in deliberation is a fast decision-maker. Phishing emails can manipulate the need for fast decision-making by creating a sense of urgency. Incident response team members need to recognise that some people will require more time and information before making a decision. Impulsivity, another aspect of personality relevant to decision-making, can also be measured, as described earlier. Alseadoon et al.[126] attributed risk-taking behaviour to personality. They tested and categorised types of phishing email victims. Risk-taking victims were found to be aware of malicious emails but perceived them to be low risk and/or felt that password protection (for a university blog) was unimportant as the information was not worth protecting. One interview transcript indicated that a risk-taker thought passwords were important for banking applications but not blogs. This highlights the need to provide training on the importance of system, information sharing, and vulnerabilities.

Scott and Bruce,[127] presented a tool to measure individual decision-making styles: General Decision-Making Style (GDMS). It incorporated five main styles: rational (logical search and analysis); intuitive (based on feelings); dependent (seeking support from others); avoidant (choosing to avoid decision-making); and spontaneous (reflecting the amount of time taken for decision-making). Raywood-Burke,[86] tested correlations between decision-making style using the GDMS, and risk-taking behaviours and preferences. Their results suggested that people who conduct rational decision-making and do not tend to avoid decision-making, were more likely to lock devices. People with these styles were also more likely to demonstrate proactive awareness and perform software updates and good password management. There was an inverse correlation between intuitive and spontaneous decision-making styles, and proactive awareness. Risk-taking was measured using the Domain-Specific Risk Taking (DOSPERT) scale.[128] The results suggested that people who take fewer ethical and recreational risks are more likely to exhibit proactive cyber security awareness. Low-risk ethical behaviour was also associated with software updating.

Group decision-making employed by a command or incident response team, can be affected by a phenomenon known as 'risky shift'. Stoner[129] identified that people make significantly more risky decisions in a group than individually. They effectively 'shift' their opinion towards that of the group. If an individual initially disagreed with the group decision, they were less confident of both their own, and the group's, decision. People are also less likely to help someone in need when in a group than if they are alone.[34] Cyber security team leaders, therefore, need to listen to the opinions of all members to avoid a decision that is led by a few dominant characters. Emotional intelligence (ET) is an important aspect of teamwork. Tsarenko and Stritzhakova[130] identified a positive correlation between ET and problem-solving and emotionally expressive coping strategies.

Analysis of individual and team decision-making and risk-taking under uncertainty can help us understand and predict how people react to a cyber security incident. Decision-making "human errors" are complex and derived from a variety of causes. The next section explores error in more detail.

HUMAN ERROR

A large proportion of cyber security incidents have been attributed to "Human error", as much as 95% in some cases.[27] This term can be misleading in that one could assume that people are to blame. Human factors professionals argue that systemic, organisational root causes are, in fact, at fault. This is discussed in Chapter 5. However, there are situations where humans are particularly vulnerable, and this can impact cyber security.

Classification of error

Unintentional error is explained by Rasmussen, 1983 (cited in Kirwan)[131] in terms of skill, rule, or knowledge-based (SRK) errors. Skill-based errors occur when someone is performing in 'autopilot' and paying little attention to the task in hand. This could cause them to inadvertently activate a malicious phishing email link or attachment. Rule-based mistakes refer to the incorrect application of plans or operating procedures. The wrong course of action is selected and/or the correct course of action is omitted. A rule appropriate for one situation may be incorrectly applied to a similar situation. For example, the social 'rule', that it is polite to hold doors open, may be inappropriate in a secure environment that is restricted to authorised personnel. Likewise, the tendency to share information may be considered appropriate on social media, but sharing personal or organisational information may create cyber security vulnerabilities. Social etiquette rules may have been a factor in the banking attack described in Chapter 2.

Knowledge-based mistakes are caused by a lack of knowledge or experience about a particular situation at a given point in time. Lack of awareness of current cyber security threats and attack vectors, and how to tackle them, may weaken an individual's understanding of the impact of their work on the resilience of the organisation. Awareness training needs to address this, and employees need to complete training provided by their organisation.

Reason, 1990 (cited in Kirwan)[131] categorised errors into slips, lapses, mistakes, and violations. Mistakes were classified as errors of intention, like inappropriate rule selection. Slips refer to the right intention carried out incorrectly, perhaps because it was mistaken for something similar, like selecting a fake website instead of the real one. A lapse is a failure to carry out an action, perhaps because of an interruption. This could account for the loss of sensitive information in laptops or paperwork.

Violations, on the other hand, are intentional behaviours. In cyber security, violations include a deliberate misuse of password or information sharing policy. Whilst these actions create vulnerability for the organisation, they are usually not malicious. They may, in fact, be performed with the intention of completing work more efficiently. If cyber security policy and procedures are too strict, people will find workarounds to them, such as transferring information using personal email accounts or unauthorised memory storage devices. Blaming users is not helpful. Instead, the procedures need to be designed around jobs and, if possible, the most secure way to perform a task should also be the easiest way. To achieve this, employees need to be consulted in the design of procedures.

Password violations arise because they are heavily reliant on limited human memory capacity. A typical user will require many passwords for personal and professional applications and websites. There is a risk that people will use the same, easy-to-remember password for several applications, so if one is compromised, the rest are too. Biometric technologies may be a better solution because they eliminate reliance on human memory. Fingerprint and facial recognition systems are common on smart phones but could be better utilised in commercial computer equipment and applications. Education can alert employees to the dangers of engaging with unauthorised websites using work equipment.

Malicious behaviour

Some violations are, of course, malicious. If employees feel unappreciated, at risk of redundancy, or disagree with an organisational policy, the risk of them compromising the organisation increases. Their situation could change since joining the organisation. They may become more susceptible to blackmail, perhaps because of financial difficulties or extramarital affairs, so initial screening mechanisms are not sufficient.

Insider threat has been categorised as intentional or unintentional.[132] An insider is someone with access to an organisation's information. The term "insider threat" could be misleading. It suggests that employees are to blame for cyber security incidents, when arguably, only intentional, malicious insiders are responsible.

Error-producing conditions

Williams[24] captured task types where people are more vulnerable, and "Error Producing Conditions" (EPC) in the Human Error Assessment and Reduction Technique (HEART). Tasks that are particularly vulnerable to error, according to the tool, include those performed at a fast pace; without supervision or procedures; high in complexity; and simple tasks performed without really thinking. Checking emails, for example, could be performed at pace without much attention. More highly complex tasks that may be performed without supervision could include software programming, updates, or installation, and writing reports.

Task performance can be affected by some EPCs more than others. One EPC is time pressure, which would increase the likelihood of an email recipient inadvertently activating a malicious link or attachment. An experiment by Williams et al.[133] found fake software update messages were more likely to be accepted when a participant was conducting a demanding short-term memory task than when more time was allowed in which to respond to the message. This suggests that time pressure and the drive for productivity can affect cyber security. Similarly, Sawyer et al.[134] found that cyber security signal detection performance declined over time, and high workload affected cyber security performance.

Tiredness is related to another EPC in HEART. It is recognised that we are more likely to make errors when we lack sleep. This could affect our concentration at work and is a particular risk factor if shift-working is not properly designed. The worst time of day for tiredness related accidents is between midnight and 6 am. The afternoon, between 2 and 6 pm is also problematic, to a lesser extent.[135] The Health and Safety Executive (HSE), provided advice about managing shiftwork, including the design of shift patterns.[136] Permanent, (non-rotating) eight-hour daytime shifts are preferred.

A lack of training and/or experience in a specific activity typically increases the probability of associated error, especially if the task is performed infrequently, according to the model. In terms of cyber security, this could mean information is shared inappropriately or using insecure platforms. Poor equipment or interface design can also give rise to errors. For example, if an application does not match the user's mental model or procedures pertaining to the task, or provides poor feedback about their inputs, errors in operation may occur. Poorly designed applications can also allow users to override security procedures. It is therefore important that design maximises usability whilst maintaining security. This balance can be difficult to achieve so human factors expertise is required to minimise the potential for Human Machine Interaction (HMI) errors. This topic is discussed in Chapter 6.

Good procedure design also helps minimise error potential. Instructions should be clear, unambiguous[137] and appropriate for the job. If employees cannot understand procedural instructions, or if they are not designed around their work, they may ignore or bypass them. This highlights the need to consult employees in the design of security procedures.

An organisational culture that encourages productivity over cyber security is likely to be less resilient. Williams captures this EPC in HEART as "an incentive to use other, more dangerous procedures". Stress and a lack of breaks from work can also induce errors. This in turn may also reduce morale and thereby further increase error potential. Insufficient numbers of security personnel are likely to cause stress. We often associate workload with stress. High workload can certainly increase the likelihood of error. However, low workload is also risky.[138] Prolonged monitoring activity can create vigilance effects; loss of concentration resulting in missed information. Security Operations Centre (SOC) network monitoring is at risk from low workload effects if not properly managed by breaks and/or variation in activity.

CONCLUSION

This chapter has explored individual differences and their impact on cyber security, as well as generic decision-making and error principles. It is arguably unethical to measure the level of vulnerability of individual employees. Just because someone exhibits high-risk traits or attributes, does not mean they will commit a crime. Furthermore, it is difficult, time-consuming, and expensive to attempt to change people. System factors are relatively easy to fix. Organisational vulnerabilities are discussed in Chapter 5.

On a positive note, employees can be an organisation's first line of defence. They have the power to identify and report phishing emails or social engineering attacks.

KEY LEARNING POINTS

Learning points from the analysis of individual vulnerabilities are summarised here and incorporated into recommendations in Chapter 6 where appropriate.

Personality

- Personality profiling may assist personal selection and training.

Age and gender

- There are apparent advantages and disadvantages of age and gender in relation to cyber security.

Culture

- Employees with a strong bond with their organisation may be more likely to react well to cyber security policy and procedures.

Decision-making

- Analysis of heuristics and cognitive biases, and their impact on cyber security decision-making, supports incident investigations and training.

Training

- Implications for incident response training and general awareness training were discussed.

Error

- Conditions such as time pressure and workload can increase the likelihood of error. Error classification theories, and understanding of known error-producing conditions, can support the prediction and mitigation of human error, and retrospective incident investigations.
- Poor design of interfaces and procedures can give rise to errors. New technologies can enhance cyber security and innovation, but they need to consider the human element in cyber security, by design.

NOTES

82 Costa, P.T. and McCrae, R.R, 2006, *Revised NEO Personality Inventory (NEO PI-R)*, Manual (UK ed.). Odessa FL: Hogrefe.

83 Widdowson, A.J., 2016, CHEAT: An updated approach for incorporating human factors in cyber security assessments, *IET Engineering Technology Reference* Vol 2016 1–7, https://doi.org/10.1049/etr.2015.0104.

84 Shappie, A.T., Dawson, C.A. and Debb, S.M., 2019, Personality as a predictor of cyber security behavior. *Psychology of Popular Media Culture.* Advance online publication. http://doi.org/10.1037/ppm0000247.

85 McBride, M., Carter, L. and Warkinten, M., 2012, Exploring the role of individual employee characteristics and personality on employee compliance with cyber security policies. (Prepared by RTI International – Institute for Homeland Security Solutions under contract 3-312-0212782).

86 Raywood-Burke, G., Bishop, L.M., Asquith, P.M. and Morgan, P.L., 2021, Human individual difference predictors in cyber-security: Exploring an alternative scale method and data resolution to modelling cyber secure behavior.

In Moallem, A. (eds) *HCI for cyber security, privacy and trust.* HCII 2021. Lecture Notes in Computer Science (vol. 12788). Cham: Springer. https://doi. org/10.1007/978-3-030-77392-2_15.

87 Egelman, S. and Peer, E., 2015, Scaling the security wall, developing a Security Behaviour Intentions Scale (SeBIS), *CHI 2015 Conference Paper*, http://dx.doi. org/10.1145/2702123.2702249.

88 Gross, M.L., Canetti, D., Dana, R. and Vashdi, D.R., 2016, The psychological effects of cyber terrorism, *Bulletin of the Atomic Scientists* 72(5), 284–291, https://doi.org/10.1080/00963402.2016.1216502.

89 Alseadoon, I., Chan, T., Foo, E. and Gonzalez Nieto, J., 2012, Who is more susceptible to phishing emails?: A Saudi Arabian study. *ACIS 2012 Proceedings*. 21. https://aisel.aisnet.org/acis2012/21.

90 Alkis, N. and Temizel, T.T., 2015, The impact of individual differences on influence strategies, *Personality and Individual Differences*, December 2015, https://doi.org/10.1016/j.paid.2015.07.037.

91 Moallem, A., 2020, Social engineering (chapter 7). In A. Moallem (Ed.), *HCI for cyber security, privacy and trust.* Boca Raton: CRC Press, HCII 2020.

92 https://www.hoganassessments.com/assessments/.

93 Lee, K. and Ashton, M.C., 2012, *The H factor of personality; why some people are manipulative, self-entitled, materialistic, and exploitive – and why it matters for everyone.* Waterloo, Ontario: Wilfrid Laurier University Press. ISBN 978-1-55458-843-3.

94 Kranenbarg, M.W., Gelder, J.-L., Barends, A.J. and de Vries, R.E., 2022, Is there a cybercriminal personality? Comparing cyber offenders and offline offenders on HEXACO personality domains and their underlying facets, *Computers in Human Behaviour* 140(2023), 107576.

95 Patterson, W. and Winston-Proctor, C.E., 2019, *Behavioral cybersecurity.* CRC Press, Boca Raton: Taylor & Francis.

96 Zaccaro, J., Dalal, R.S., Tetrick, L.E. and Steinke, J.A., (Eds.) 2016, *Psychosocial dynamics of cyber security.* Abingdon Oxon and New York: Routledge.

97 Safa, N.S., Von Solms, R. and Furnell, S., 2016, Information security policy compliance model in organisations, Computers and Security Vol 56, 70–82, *Science Direct.*

98 Katja Gelbrich, Jana Gäthke, Yany Grégoire, 2016, How a firm's best versus normal customers react to compensation after a service failure, *Journal of Business Research* 69(10), 4331–4339, ISSN 0148-2963, https://doi.org/ 10.1016/j.jbusres.2016.04.010.

99 Van de Weijer, S.G.A. and Moneva, A., 2022, Familial concentration of crime in a digital era: Criminal behaviour among members of cyber offenders, *Computers in Human Behavior Reports* 8(2022), 100249.

100 Suler, J., 2004, The online disinhibition effect, *Cyberpsychology & Behavior* 7(3), 2004.

101 Morgan, P.L., Soteriou, R., Williams, C. and Zhang, Q. (2020). Attempting to reduce susceptibility to fraudulent computer Pop-Ups using Malevolence cue identification training. In T. Ahram & W. Karwowski (Eds.), *Advances in Human Factors in Cyber security.* AHFE 2019. Advances in Intelligent Systems and Computing (vol. 960). Cham: Springer. https://doi.org/10.1007/9 78-3-030-20488-4_1.

102　Vishwanath, A., Harrison, B. and Ng, Y.J., 2018, Suspicion, cognition, and automaticity model of phishing susceptibility. *Communication Research* 45(8), 1146–1166, https://doi.org/10.1177/0093650215627483.

103　Festinger, L., 1957, *A theory of cognitive dissonance*, California: Stanford University Press, ISBN 0-8047-0911-4.

104　Noam Ben-Asher, N. and Gonzalez, C., 2015, Effects of cyber security knowledge on attack detection, *Computers in Human Behavior* 48, 51–61, ISSN 0747-5632, https://doi.org/10.1016/j.chb.2015.01.039.

105　Wright, R.T. and Marett, K., 2010, The influence of experiential and dispositional factors in phishing: An empirical investigation of the deceived, *Journal of Management Information Systems* 27(1), 273–303, https://doi.org/10.2753/MIS0742-1222270111.

106　Bessier, K., Ceaparu, I., Lazar, J., Robinson, J. and Shneiderman, B., 2002, Understanding computer frustration: Measuring and modeling the disruption from poor designs, Technical Reports from UMIACS, http://drum.lib.umd.edu/handle/1903/1233.

107　Morgan, L.P., Williams, E.J., Zook, N.A. and Christopher, G., 2018, Exploring older adult susceptibility to fraudulent computer pop-up interruptions. In T. Z. Ahram & D. Nicholson (Eds.), *Advances in human factors in cyber security: Proceedings of the AHFE 2018 International Conference on Human Factors in Cyber security*, July 21–25, 2018, Loews Sapphire Falls Resort at Universal Studios, Orlando, Florida, USA (pp. 56–68). (Advances in Intelligent Systems and Computing; vol. 782). Cham: Springer. https://doi.org/10.1007/978-3-319-94782-2_6.

108　Branley-Bell, D., Lynne Coventry, Dixon, M., Joinson, A. and Briggs, P., 2022, Exploring age and gender differences in ICT Cyber security behaviour, *Human Behavior and Emerging Technologies* 2022, Article ID 2693080, 10, https://doi.org/10.1155/2022/2693080.

109　Henley, M., Dean, T., Schroeder, W., Houy, M. Trzeciak, R.F. and Montelibano, J., 2011, *An analysis of technical observations in insider theft of intellectual property cases*, Technical Note CMU/SEI-2011-TN-006, Pittsburgh: Carneigie Mellon University.

110　Cappelli, D., Moore, A.P., Trzeciak, R.F. and Shimeall, T., 2006, *Common sense guide to prevention and detection of insider threat*, version 3.1, CyLab, Carnegie Mellon University.

111　Kato,Y. and Kato, S., 2015, Reply speed to mobile text messages among Japanese college students: When a quick reply is preferred and a late reply is acceptable, *Computers in Human Behavior* 44, 209–219, ISSN 0747-5632, https://doi.org/10.1016/j.chb.2014.11.047.

112　Wickens, C.D, 1992, *Engineering psychology and human performance* (2nd ed.). New York: Harper Collins. ISBN 0-673-46161-0.

113　Tversky, A. and Kahneman, D. (1974). Judgment under uncertainty: Heuristics and biases, *Science* 185(4157), 1124–1131, http://doi.org/10.1126/science.185.4157.1124.

114　Transportation Safety Board of Canada, 2014, Aviation investigation report A11H002, Controlled Flight into terrain, Bradley Air Servces Limited (First Air) Boeing 737-210C, C-GNWN, Resolute Bay, Nunavut 20 August 2011.

115 Saposnik, G., Redelmeier, D., Ruff, C.C. and Tobler, P.N., 2016, Cognitive biases associated with medical decisions, *BMC Medical Informatics and Decision Making* 16, Article number 138.

116 Saks, M.J., Risinger, D.M., Rosenthal, R. and Thomspon, W.C., 2023, Context effects in forensic science: A review and application of science to crime laboratory practice in the United States, *Science & Justice* 43(2), 77–90, http://doi.org/10.1016/S1355-0306(03)71747-X

117 Wallsten, T. S. and Barton, C., 1982, Processing probabilistic multidimensional information for decisions, *Journal of Experimental Psychology: Learning, Memory, and Cognition* 8(5), 361–384, https://doi.org/10.1037/0278-7393.8.5.361

118 Ehrlinger, J. and Kim, B., 2016, Decision-making and cognitive bias, https://doi.org/10.1016/B978-0-12-397045-9.00206-8

119 Kahneman, D., Knetcg, J.L. and Thaler, R.H., 1990, Experimental tests of the endowment effect and the Coase theorem, *Journal of Political Economy* 98(6), 1325–48.

120 Ting, D., 2022, *Why cognitive biases and heuristics lead to an under-investment in cyber security.* Medford MA: Tufts University. https://www.cs.tufts.edu/comp/116/archive/fall2019/dting.pdf

121 Murata, A. et al., 2015, Influence of cognitive biases in distorting decision-making and leading to critical unfavourable incidents, *Safety* 1, 44–58, https://doi.org/10.3390/safety1010044

122 Alnifie, K.M. and Kim, C., 2023, Appraising the manifestation of optimism bias and its impact on human perception of cyber security: A meta analysis, *Journal of Information Security* 14(2), https://doi.org/10.4236/jis.2023.142007

123 Carpenter, S., Zhu, F. and Kolimi, S., 2014, Reducing online identity disclosure using warnings, *Applied Ergonomics* 45(2014), 1337–1342.

124 Kahneman, D. and Tversky, A., 1984, Choices, values, and frames, *American Psychologist* 39(4), 341–350, https://doi.org/10.1037/0003-066X.39.4.341

125 Imgraben, J., Engelbrecht, A. and Choo, K.-K.R., 2014, Always connected, but are smart mobile users getting more security savvy? A survey of smart mobile device users, *Behaviour & Information Technology* 33(12), 1347–1360, https://doi.org/10.1080/0144929X.2014.934286

126 Alseadoon, I.M., Othman, M.F.I., Foo, E. and Chan, T., 2013, Typology of phishing email victims based on their behavioural response, *Proceedings of the Nineteenth Americas Conference on Information Systems*, Chicago, Illinois, August 15–17, 2013.

127 Scott, S.G. and Bruce, R.A., 1995, Decision-making style: The development and assessment of a new measure, *Educational and Psychological Measurement* Vol. 55, 818–831.

128 Blais, A.R. and Weber, E.U. 2006. A domain-specific risk-taking (DOSPERT) scale for adult populations, *Judgement and Decision Making* 1(1), 33–47.

129 Stoner, J.A.F., 1961, *A comparison of individual and group decisions involving risk.* Cambridge: Massachusetts Institute of Technology (MIT).

130 Tsarenko, Y. and Strizhakova, Y., 2013, Coping with service failures: The role of emotional intelligence, self-efficacy and intention to complain, *European Journal of Marketing* 47(1/2), 71–92, https://doi.org/10.1108/03090561311285466

131 Kirwan, B., 1994, *A guide to practical human reliability assessment*, Boca Raton FL: Taylor & Francis Ltd. ISBN 0-7484-0111-3.
132 Cyber security & Infrastructure Security Agency, America (access 2024): https://www.cisa.gov/topics/physical-security/insider-threat-mitigation/defining-insider-threats.
133 Williams, E.J., Morgan, P.L. and Joinson, A.N., 2017, Press accept to update now: Individual differences in susceptibility to malevolent interruptions, *Decision Support Systems* 96(2017), 119–129.
134 Sawyer, B.D., Finomore, V.S., Funke, G.J., Mancuso, V.F., Funke, M.E., Matthews, G. and Warm, J.S., 2014, Cyber vigilance: Effects of signal probability and event rate, *Proceedings of the Human Factors and Ergonomics Society Annual Meeting* 58(1), 1771–1775, https://doi.org/10.1177/1541931214581369.
135 Office of Rail Regulation, 2006, Managing fatigue in safety critical work: Railways and Other Guided Transport Systems (Safety) Regulations.
136 Health and Safety Executive, 2006, Managing shiftwork, Health and safety guidance, HSG256, ISBN 978 0 7176 6197 8.
137 Bettman, J.R., Payne, J.W. and Stalin, R., 1986, Cognitive considerations in designing effective labels for presenting risk information. *Journal of Public Policy & Marketing* 5(1), 1–28, https://doi.org/10.1177/074391568600500101.
138 Yerkes, R.M. and Dodson, J.D., 1908, The relation of strength of stimulus to rapidity of habit-formation, *Journal of Comparative Neurology and Psychology* 18(5), 459–482.

Organisational vulnerabilities

5

Whilst individual employee behaviour or 'human error' is blamed for many cyber security incidents, as discussed in the previous chapters, organisational root causes are usually involved.[139] The interaction between employees and the components of their organisation affects the cyber security and resilience of that organisation. An added benefit of considering system and organisational vulnerabilities is that they tend to be easier to measure and change than human behaviour. Developing resilience ultimately needs commitment from the top. Unfortunately, Gartner[140] found under 10% of boards of directors had a dedicated cyber security committee. Furthermore, Burg et al.[141] reported that only 9% of boards were 'extremely confident' that their risk and mitigation measures protected them from significant cyber-attacks. Insufficient investment in cyber security was also found to be an issue. This chapter explores human factors-related organisational cyber security vulnerabilities. Solutions are presented in Chapter 6.

'Organisational culture' refers to shared beliefs, attitudes, and perceptions about how to behave in an organisation. An organisation's culture may be comprised of several sub-cultures.[48] For example, cyber security personnel may have a different culture to marketing personnel; senior managers may have different perceptions about acceptable behaviour compared to more junior workers; and an innovation department may have different beliefs about ways of working than a legal department. Some types of organisational culture, from the OCAI,[64] were discussed in terms of motivation and implications for cyber security (see Chapter 3). Other examples of culture-type classification were described by McKenna and created by Harrison, 1972 and Deal & Kennedy, 1982. The theories generally agree. In summary, the high-level classifications include bureaucratic culture (with lots of processes and procedures); people-oriented culture (friendly, sociable, and supportive); competitive culture (focused on achievements and results); and innovative culture (creative and less focused on rule-following).

Each culture type has different implications for cyber security behaviours. Employees in more bureaucratic organisations may be more accepting of cyber security policies and procedures because they are accustomed to procedural compliance. However, the danger is that they might feel overwhelmed with too many procedures, and new ones could get lost in the noise. People-oriented cultures may be more likely to provide good emotional support mechanisms and recognise any potential deviant behaviour. As described in Chapter 4 (individual vulnerabilities), agreeable personality types (such as those that might be found in a people-orientated culture) were more willing to comply with cyber security procedures. However, on the flip side, people in this type of organisation may be more likely to chat about sensitive information, share computer login information or hold doors open. People in innovative cultures may be more resistant to rigid cyber security procedures.

Competitive organisations may need to work with external parties under time pressure and share product information in marketing campaigns. These activities could potentially create cyber vulnerabilities. A culture where employees believe that productivity and winning business are valued more highly than cyber security would encourage potentially unsafe procedural workarounds. If the workarounds are ignored and there are no repercussions, they are likely to become an accepted norm.[142] Time pressure is also known to increase the likelihood of errors,[24] for example when sharing information by email. Employees in competitive organisations could also be influenced by the cultures of external organisations.

In an ideal culture, employees would recognise the importance of cyber security and take ownership of it in their personal and professional lives. They would feel and able to report incidents without fear of blame or punishment, a 'just culture'.[51] The position of the Chief Information Security Officer (CISO) and another security personnel is indicative of an organisation's culture and attitude towards cyber security. Culture transformation programmes are described in Chapter 6 as part of the solution. Behaviours and organisational practices that affect cyber resilience are addressed throughout this chapter.

MORALE

Morale can affect organisational culture and attitudes to cyber security. Good managers are aware of the morale of their employees. There are ways of measuring morale such as job satisfaction or engagement surveys. Prybil,[143] for example, determined a way of measuring job satisfaction which focused on prestige and the opportunity to participate in determining work methods and procedures. Targeting the sources of low morale is essential to reduce the risk of unsafe cyber security behaviours. Employees who lack a sense of autonomy, and are unable to make their own decisions freely, may abuse their access to privileged information as a way of feeling more powerful. Rogojan[144] found employees with an external locus of control, in other words, people who feel that the outcomes of their behaviours are determined by external factors rather than themselves, were more likely to engage in deviant behaviour. Deviant behaviour could simply manifest itself as a rejection of cyber security procedures, but it could be more serious and give rise to a malicious insider attack. These findings provide strong evidence for the participative design of cyber security procedures, where employees have the opportunity to provide feedback and influence the design of the procedures they are expected to follow.

In addition to a perceived lack of autonomy, the extent to which employees feel appreciated can influence their motivation and behaviour (Wong et al., 2019 cited in Triplett).[145] If people feel unappreciated, perhaps because of failure to get promoted, lack of positive feedback, or the threat of redundancy, the risk of them compromising the organisation increases.

POLICIES AND PROCEDURES

Policies and procedures help prepare employees and organisations for cyber security incidents. They may be preventative, in the case of information management, or they may guide incident response and recovery. However, if cyber security policy and the associated procedures are not designed around the people who are expected to use them, they become largely ineffectual, likely to be followed by only the most conscientious, agreeable personnel.[142] Other people will find workarounds or short-cuts to the procedures. This behaviour is not usually malicious. They are usually just trying to find the most efficient way of doing their job. For example, if employees need to share information with third-party external organisations, and the procedures prevent them from doing this, they may resort to unsafe methods such as using their personal email account or unauthorised peripheral devices such as USB memory sticks. While monitoring of information sharing may help, it is also necessary to ensure that procedures are designed around the employees' jobs. To do that, employees need to be consulted in a participatory design process. Potential workarounds should be anticipated so that safer alternatives can be implemented. If possible, the most secure way to perform a task should also be the easiest way.

User validation policies that require employees to use lengthy complex passwords and change them frequently could increase the likelihood of risky cyber security behaviours such as writing passwords down, storing them in a mobile phone, or on sticky notes; reusing them, sharing them, or using easy to remember (and therefore guess) passwords.

To facilitate compliance, people also need to be able to *access* the policies and procedures and know where to find them. If they are badly written, lack brevity, and are difficult to find, employees may be less likely to adopt them. The title of the procedure is also important. Using a mix of terms, such as 'cyber security', 'information management' or 'acceptable use', may be confusing to employees. It may not be immediately obvious to all employees that those policies refer to cyber security.

In addition to consistent labelling of policy and procedures, official communications need to comply. For example, if a cyber security policy prohibits clicking on links in emails, and the internal communications department routinely sends emails inviting employees to click links, the policy is contradicted and the behaviour, (in this case, clicking on links), becomes an accepted norm. Policy and behaviours therefore need to be led from the top and integrated throughout the organisation. If a person is observed routinely accepting pop-up messages without reading them, and this is not questioned, then the behaviour becomes a norm and is likely to continue.

RECRUITMENT, SELECTION, AND STAFFING

Recruitment is an opportunity to identify any vulnerabilities that potential employees may have. Screening of candidates is commonplace where sensitive government information is held. However, it is important to remember that people's circumstances can change over time. They may develop financial difficulties or marital problems that increase their susceptibility to blackmail or coercion from malicious parties, for example.

When we interview and select people to join an organisation, some traditionally valued behaviours and personality traits are not necessarily optimal for cyber security. For example, highly agreeable, social individuals may be great to work with, but, as described earlier, they may create vulnerabilities in terms of data protection. It is therefore necessary to identify a balance between traits that are desirable for a particular job, and cyber security. This is especially important for security personnel and roles with access to sensitive information.

The HR department can be an easy target for malicious outsiders. They routinely expected to receive legitimate unsolicited CVs and applications from external personnel. So, it would be relatively easy for a malicious outsider to send electronic information that incorporates malware, under the guise of a job application. Other departments are also at risk. For example, technical and legal teams process commercially sensitive information, which could damage the organisation if lost. The marketing department may inadvertently share sensitive information while promoting the organisation in the public domain.

THIRD PARTIES

Organisations routinely interact with third parties in the course of their operations. They include contractors, suppliers, and customers. All these groups potentially provide another route to attacking the organisation. If a third-party organisation is compromised, so is the information that they hold about the organisation. Therefore, it is important to ensure that associated third parties are aware of the organisation's cyber security policy and procedures, and that they take the necessary steps to protect themselves, and therefore the organisation.

EDUCATION AND AWARENESS

Cyber security education and awareness training have often been seen as a solution to the 'problematic' human element in cyber security. Whilst human factors experts would consider organisational or systemic root causes when determining vulnerabilities, training is nevertheless one important factor.[146] Organisational surveys have suggested employees can lack awareness of cyber security and their organisation's cyber security policy. This could be caused by difficult to access, or inappropriate, security policies and procedures, but it could also be because of poorly designed training. Poor uptake of cyber security training leaves an organisation vulnerable. It is therefore necessary to consider the reasons why employees might ignore or fail to understand cyber security training that is provided.

One possible reason employees might ignore cyber security training is that they simply don't see it as important enough, especially considering their other pressing work commitments, a culture issue. Cyber security training will likely have to compete with other training expected by the organisation.

Another reason why training is either ignored or not understood is when it is just not very good. If it does not contain examples that are recent and relevant to the target audience, they may find it hard to relate to it. They may consider a cyber security incident as something that happens to other people, especially if they themselves have an external locus of control. The threat is therefore downplayed. Or perhaps the training is ignored because people think that somebody else, like their IT engineer(s), would protect them from cyber-attacks. They may have heard of data protection legislation, such as GDPR,[2] but assume it is somebody else's responsibility and does not affect them. The training may not address the trainees' intrinsic motivations. Different personalities could respond to the threats in different ways (For more details on personality and cyber security, see Chapter 4).

People may be unaware of the frequency and volume of attacks that their organisation receives, so they are detached from the severity of the threat. Alternatively, if they are aware of the threat, they may not appreciate the potential consequences for their organisation. The organisation itself may not be fully aware of all threats, of course. If incidents are not reported, for example because of fear of blame and punishment, they are unlikely to find their way into awareness communications and training. Smaller organisations might not have a formal CISO position and may consequently lack awareness of how to provide the necessary training.

If there is poor understanding of how well employees have understood the training, success cannot be assumed. Some training courses simply provide multiple-choice questions with fairly obvious answers. Employees will be tempted to simply click through the training and multiple-choice options without really internalising the desired message.

The timing of training is also relevant. If it is not part of induction training, new recruits may find this indicative of the organisational culture and a poor attitude towards cyber security. If it is *only* provided during induction, it may be forgotten over time. If the training takes place and nothing bad happens, the message may be undermined, hence the need to keep training current by including recent relevant attack examples. Frequent training may help maintain the message. However, employees can become overloaded or flooded with information and therefore forget or ignore some of it.

In addition to preventative training and awareness, employees also need to know what to do when an incident happens. Some may not know how to report an incident such as a phishing email attack, or they might just find it too difficult and consider it a waste of time. People react differently in a stressful incident scenario, and their decision-making is affected, as discussed in Chapter 4. Senior managers may not know what to do, or where to seek information, when they fall victim to a serious attack. Furthermore, the incident response could be jeopardised if a few dominant characters were allowed too much influence, and others found it difficult to get their concerns heard. An incident response plan and associated training are therefore necessary to mitigate these risks.

TECHNOLOGY

Technological vulnerabilities tend to arise from either organisational issues or misuse of technology by individual employees. To solve the latter, it is necessary to address the former. Organisational issues can stem from an overall lack of investment in cyber security. Fernandez De Arroyabe et al.[147] found that organisational investment in cyber security was reactive. More cyber security-capable organisations invested a greater amount. It is dangerous to wait for a severe attack before investing. One of the affected targets of the 2017 global "Wannacry" ransomware attack, the National Health Service (NHS) in the UK, was vulnerable because of such a lack of investment. Healthcare provision was temporarily compromised because of the attack. Microsoft issued a patch to address the vulnerability prior to the attack, but it did not cover the older operating system used by affected parts of the NHS. An emergency patch was later released to cover these systems. Prior investment in more recent operating systems and regular patching should have provided resilience against this attack.

A comprehensive asset management system would identify which machines use which operating system; close redundant accounts; and identify when patches have been deployed. Such a system is, itself, reliant on investment in cyber security and sufficient numbers of competent, trained maintenance personnel. This could be a particular problem for smaller organisations with no dedicated Information Security (IS) team. Malicious actors target systems that have not invested in updated antivirus protection.

Poor ergonomic design of technology increases the probability of human error and, therefore, reduces organisational resilience. Thron et al.[148] identified that rail signallers found it difficult to differentiate between genuine system faults and a cyber-attack, so the design needs to recognise this and seek to assist the user. Monitoring security systems for unusual behaviour relies on human attentional capabilities. If users become overwhelmed with alarms because of a bad interface design or are distracted by other events, they are likely to miss information, so the design of alarm management systems is important.

Presentation of cyber security information is another important factor. Information needs to be conveyed in a way that the target audience will understand. Otherwise, the threats may be ignored and required risk mitigation strategies could fail to materialise. Too much information may be unnecessary and overwhelm some people, given the amount of other information they are required to process. Colour-coding can be used to make sense of information. However, user colour-coding conventions vary so it cannot be assumed that a given colour will be interpreted in the same way by everyone. Certain combinations of colours (such as bright red and blue) can cause unwanted visual effects and eyestrain, and others are difficult for colour-deficient users to distinguish.[149]

Like, colour-coding, blink-coding can affect usability. Movement in our peripheral vision is distracting. Auditory alerts are also disruptive. Flashing alerts and auditory alerts should only be used to attract attention. Icons can be used to direct attention but lose their usefulness if they are not understood by the target audience. According to Gestalt theory of perception, on-screen items that are grouped in close proximity or similarity are interpreted as being related to each other.[150] Mistakes can occur when an incorrect item is selected instead of a similar (correct) item nearby. Excessive, ungrouped screen clutter makes it harder for users to find the information they need. Gestalt theory also tells us that the brain tends to interpret incomplete shapes as complete, which has implications for pattern recognition and the design of displays for monitoring abnormal network behaviour. Human factors advice regarding presentation of information and the user experience (UX) of cyber security is discussed in Chapter 6.

An example of employee misuse of technology is the activation of free public Wi-Fi, perhaps where they have a poor phone signal. Expectancy theory (see Chapter 3) helps explain this. The desire for immediate gratification and the need to complete Internet-based activities can override security concerns. Virtual Private Networks (VPN) encrypt networked information. Failing to administer and mandate the use of a VPN would mean employees' work may be compromised by attacks.

Internet misuse is another employee-related risk. People could inadvertently provide information on fake websites or download infected software. They can share information about the organisation on social media and generative AI sites. This could be done with the best of intentions, such as promoting achievements or enhancing product development. However, some information could potentially compromise the organisation. Furthermore, attackers can trick generative AI tools into revealing passwords.[151] Misuse of email and USB devices, perhaps exacerbated by overly strict information-sharing restrictions, as discussed earlier, could similarly result in a loss of sensitive information. The large volume of emails received by employees, especially those in senior positions or with a wide network, means they may be managed in a rush. Emails may not be always given the recipient's full attention, so phishing emails may be missed. Information-sharing vulnerabilities can also occur when using video-conferencing platforms.

The use of Personal Electronic Devices (PED) for work emails and files restricts the ability of the organisation to protect its data. Emails sent using PEDs are outside the control of protective mechanisms such as a VPN and therefore at risk of exposure. Strong passwords cannot be enforced; patching and antivirus protection is reliant on individuals; unauthorised USB devices can be used to transmit malware to work systems. Unencrypted USBs, laptops, and phones do not protect information when they are lost or stolen. A 2022 UK Government Cyber Breaches Survey[6] found that the use of personal devices (such as smartphones) by employees is, unfortunately, common. A worrying number of laptops and other mobile devices containing sensitive information are lost or stolen. Over 2,000 devices were lost or stolen from UK Government employees in a year, according to reports including the BBC.[152] This highlights the need for device encryption to protect against human vulnerabilities like memory lapses.

Other employee technology-related vulnerabilities are caused by limited human memory capacity. Employees may forget to lock their computer when they leave their desk. They may forget to collect sensitive information on printers or desks. They may inadvertently leave a laptop on a train or in a cafe. Organisations need to find solutions for these vulnerabilities.

Password management was mentioned in the policy section earlier in this chapter. The use of passwords for user validation is heavily reliant on limited human memory capacity. A typical user will require many passwords for personal and professional applications and websites. There is a risk that people will use the same, easy-to-remember password for several applications, so if one is compromised, the rest are too.

In summary, technological advancements can enhance our way of life, but they also bring additional susceptibilities. Autonomous vehicles, for example, may eventually decrease the risk of human error and fatigue by reducing the reliance on the human driver. However, if they are hacked, attackers may gain control of national infrastructure components such as traffic lights. In an ideal world, all employees would take ownership of cyber security when using technology, rather than attributing attacks to luck or assuming someone else has the responsibility to protect them. Unfortunately, the work of Klebl[153] suggests this is not the case. They found that people who thought that luck determined errors when using technology, experienced more technical difficulties than those who felt competent in their ability to work with it. Organisations therefore need to equip their employees with the skills to manage technology. It is not all bad news, however. In addition to creating vulnerabilities, technology can improve cyber security. Mosteanu[154] found that, with the appropriate skills and integration of operations and IT departments, new technologies and cyber defence systems can support organisations in the fight against cybercrime.

PHYSICAL WORKING ENVIRONMENT

Although it may not sound relevant to cyber security, the physical working environment is a cyber security vulnerability. If malicious outsiders gain access to a building, they can also access any sensitive information contained within, unless suitable precautions are taken. There are several ways in which human vulnerabilities contribute to the resilience of the working environment. Holding doors open is one such method. Attackers rely on the social construct of politeness to enable them to access buildings and will typically target nervous-looking individuals they think are least likely to challenge them. Attack strategies might include walking on crutches or holding a cup of coffee in each of their hands to encourage unassuming employees to open doors for them. Another tactic might be dropping a pen in the door jam, so when someone walks through, the door will not close.

People can feel awkward about asking to see someone's credentials. They don't want to appear rude and in breach of the unwritten rules of social compliance. Attackers can capitalise on this. Turnstiles may help but are also vulnerable to tailgating, where unauthorised personnel follow a legitimate employee through. Social pressures can deter employees from confronting such behaviour. An attacker may carry a fake authorisation letter as an extra layer of protection. This means that even if they are challenged and the employee calls the verification number on the letter, the employee unknowingly talks to a criminal accomplice, rather than an authorised person within their company.

Once inside the building, hard copy information left on desks or printers can be seen or obtained. Electronic information can be extracted or manipulated through unlocked/unprotected computers and server rooms. Operational Technology (OT) environments, such as found in nuclear power plants, were traditionally secure because computer devices were 'air-gapped' (not connected to an external network). This would not protect them from a physical environment breach, however. In this situation, attackers can gain access to electronic systems by inserting a USB device into an unlocked machine. They can either insert the device themselves through gaining unauthorised access to a building or they can leave an infected device for an unassuming employee to insert. A tiny, inexpensive computer, such as a Raspberry Pi, can also be installed to enable attackers to access information remotely.

If people are used to seeing strangers in their working environment, they might be less likely to recognise and challenge an unauthorised criminal. Attackers may try to dress like others in the organisation in an attempt to fit in. Even if an employee identifies a suspicious person, diffusion of responsibility[34] may prevent them from approaching or reporting them.

Office environments are not the only physical environments where an organisation's information is vulnerable. In public environments, if recognised, an organisation's employees may inadvertently reveal sensitive information. In cafes, pubs, or public transport, for example, they may chat about their work and thereby provide malicious personnel in the vicinity with information that could support an attack. Unauthorised information can be extracted by engaging with individuals or simply eavesdropping on their conversations. Clothing that is branded with the organisation's logo may be encouraged from a marketing perspective. However, if it is worn in public, it could be used by unscrupulous outsiders to target employees who have access to proprietary information.

Shared office spaces may be a particular issue for small companies. The building infrastructure may not allow their employees the privacy needed to discuss sensitive information without being overheard by unauthorised parties. The rise of remote and home working creates additional vulnerability in this respect. Information shared on a video-conferencing call is at risk of being overheard by family members or other household occupants. Hard copy information or unlocked computers are also vulnerable in these sorts of environments. Delbosc et al.[155] found employer and colleagues' attitudes, towards working from home, were more influential to employees than whether they felt capable of doing so. Employers therefore need to ensure that remote-working employees have the capability in the form of secure equipment and environment, if they need to access sensitive information.

It is clear that physical environment cannot be ignored.

INCIDENT MANAGEMENT

How organisations manage incidents when they happen is key to their cyber resilience. An incident could be a social engineering approach; inadvertent activation of malware; or provision of login details to fake websites (see Chapter 2 for more incident examples). Employees may be deterred from reporting incidents for several reasons. They may fear blame and punishment,[51] or they may not know *how* to report an incident and lack the time or inclination to find out. Onerous reporting processes may also affect reporting. It is human nature for people to find the easiest way of doing their job, so if reporting an incident is perceived to be too hard or takes too much time, they may not bother. This would mean organisations would not get sight of the incident.

Once incidents are known, it is not enough to simply fix the immediate problem. It is also necessary to consider the root causes. A lack of consideration of human vulnerabilities, such as those described in this book, would reduce the accuracy of any investigation findings. Participation in incident investigations has unfortunately seen training recommended for the individuals concerned in the incident, without proper consideration of the organisational failures. Training is not the only solution. It can be hard to obtain root causes of incidents because companies can be reluctant to share them. This limits their ability to learn from each other. If employees are unaware of the frequency, volume, and severity of cyber-attacks on their organisation, they may not appreciate the level of threat and the need to follow procedures and perform good cyber security behaviours.

Some incidents are relatively harmless because they are protected by control mechanisms. Others are more impactful. Senior managers need to know how to respond to a significant successful cyber-attack to minimise damage. Management/security team-working behaviours and emotional responses can affect incident response, as discussed in Chapter 4. Poor situational awareness could cause managers to deal with the immediate threat at the expense of the bigger picture. Incident response procedures and training are therefore important.

CONCLUSION

An understanding of factors that contribute to cyber security incidents can really help organisations enhance their resilience. Reason's "Swiss cheese model"[139] illustrates that systemic, organisational factors need to be considered, rather than solely focusing on the individuals involved. Several factors have been discussed in this chapter. Culture affects the organisation's attitude towards cyber security and associated ways of working. Overly strict procedures are at risk of being ignored, so workers need to be consulted in the design of procedures that affect them. Third parties need to be scrutinised and made aware of those procedures. The recruitment process may employ tailored selection measures, but personality traits suitable for one role may be unsuitable for cyber security roles. Cyber security needs to be recognised at a senior level, and associated board reports should be designed around the target audience. The organisation needs to know how to tailor training, and to what extent staff members have taken and understood it. Opportunities for misuse of technology need to be addressed. There needs to be sufficient investment in technology to support cyber security, and human factors interface design considerations should be implemented. The physical working environment, whilst not immediately obvious to cyber security, also poses risks for sensitive information. Finally, the way the organisation manages and shares incidents is critical to its overall cyber security resilience.

These factors should not be considered in isolation. The interaction between them affects cyber resilience. For example, incident reporting and investment in cyber security can be affected by the organisational culture. Similarly, failure to complete cyber security training could result in misuse of technology. Situational factors can also have an impact, as observed during the COVID-19 pandemic, where employees were forced to work from home where possible. Solutions to address organisational factors, as well as motivation and individual differences, are proposed in Chapter 6.

KEY LEARNING POINTS

Learning points have been categorised and incorporated into the recommendations in Chapter 6 where appropriate.

General

- Organisations lack confidence in their ability to address cyber security issues.
- Systemic, organisational factors need to be considered, rather than solely focusing on the individuals involved.

Culture

- Types of organisational culture have different implications for cyber security behaviours.
- Morale can affect culture and attitudes to cyber security so needs to be analysed and measured.

Training

- Organisational culture can affect attitude towards training.
- Training is not the only solution to human-related cyber security incidents.

Procedures

- Employees are likely to find workarounds to overly strict procedures.
- Third parties can create vulnerability.

NOTES

139 Reason J., 2000, Human error: Models and management, *BMJ* 320(7237), 768–770. https://doi.org/10.1136/bmj.320.7237.768. PMID: 10720363; PMCID: PMC1117770.

140 Gartner, 2021, Gartner predicts 40% of boards will have a dedicated cyber security committee by 2025. https://www.gartner.com/en/newsroom/press-releases/2021-01-28-gartner-predicts-40--of-boards-will-have-a-dedicated-

141 Burg, D., Hussain, A. and Watson, R., 2021, Cyber security: How to you risk above the waves of a perfect storm? The EY Global Information Security Survey 2021 finds CISOs and security leaders battling a new wave of threats unleashed by COVID-19, https://www.ey.com/en_gl/cybersecurity/cybersecurity-how-do-you-rise-above-the-waves-of-a-perfect-storm

142 Emma, W., 2017, Growing positive security cultures, https://www.ncsc.gov.uk/blog-post/growing-positive-security-cultures.

143 Prybil, L.D., 1973, Job satisfaction in relation to job performance and occupation, *Personnel Journal* 52(2), 94–100.

144 Rogojan. P.-T., 2009, *Deviant workplace behavior in organizations: Antecedents, influences, and remedies*, Vienna: Universität Wien.

145 Triplett, W.J., 2022, Addressing human factors in cyber security leadership, *Journal of Cyber security and Privacy* 2(3), 573–586, Special issue Cyber Situational Awareness Techniques and Human Factors. https://doi.org/10.3390/jcp2030029

146 Alotaibi, M., Furnell, S. and Clarke, N., 2016, Information security policies: A review of challenges and influencing factors, *11th International Conference for Internet Technology and Secured Transactions* (ICITST-2016) Barcelona, Spain.

147 Fernandez De Arroyabe, I., Arranz, C.F.A., Arroyabe, M.F. and Fernandez De Arroyabe, J.C., 2023, Cyber security capabilities and cyber-attacks as drivers of investment in cyber security systems: A survey for 2018 and 2019, *Computers and Security* 124, 102954.

148 Thron, E., Faily, S., Dogan, H. and Freer, M., 2024, Human factors and cyber-security risks on the railway – the critical role played by signalling operations, *Information and Computer Security* 32(2), 236–263, Emerald Publishing Limited.

149 BS EN ISO9241-112 BSI, Ergonomics of Human-System Interaction (2017) Part 112; Presentation of Information.

150 Mohamed, K. and Hussein, A., 2023, Analysing the role of Gestalt elements and design principles in logo and branding, *International Journal of Humanities and Social Science*, 10(2), 1–11. https://doi.org/10.14445/2349641X/IJCMS-V10I2P104

151 Blythe, J., Breen, K. and Iqbal, J., 2024, *Unveiling the dark side of GenAI: How people trick bots into revealing company secrets*, Bristol: Immersive labs.

152 BBC, 20th February 2020, Thousands of mobile devices and laptops lost by UK government in a year. https://www.bbc.co.uk/news/technology-51572578

153 Klebl, M., 2014, *Dealing with malfunction: Locus of control in web-conferencing, International Conference e-Learning*, Portugal.

154 Mosteanu, N.R., 2020, Challenges for organisational structure and design as a result of digitalisation and cyber security, *The Business and Management Review* 11(1), 278–286.

155 Delbosc, A., Currie, G., Jain, T. and Aston, L., 2022, The 're-norming' of working from home during COVID-19: A transtheoretical behaviour change model of a major unplanned disruption, *Transport Policy* 127(2022), 15–21.

Mitigating solutions

6

Earlier chapters have described the motivation behind human behaviour pertaining to cyber security, and individual and organisational vulnerabilities. Solutions are now proposed.

CONDUCT RISK AND VULNERABILITY ASSESSMENT

Executive boards need to understand how to manage cyber security. That means they need to know what is required of them and be able to receive clear advice from competent security personnel. The CISO's reporting line is one consideration. Board members need to understand the threats if they are to invest in cyber security. Investment means providing adequate resources, equipment, and security personnel. It means conducting risk assessments and implementing associated recommendations. One example of a governance model to manage (cyber security) risk is the IIA's three lines of defence.[156] The first line is focused on the management of risk. The second line encompasses monitoring and challenging risks. The third line provides internal, independent assurance and advice. Risk management is the first of the UK National Cyber Security Centre's (NCSC) "10 steps to cyber security".[157] It should be an ongoing process. The NCSC toolkit for boards[158] also advocates risk assessments and regular reviews, as well as the identification of critical assets to support risk management, and threat identification and evaluation. Given the high proportion of cyber security incidents attributed to "human error",[27] it is essential to consider the impact of threats on employees and their potential contribution to incidents. Employee questionnaires are one method of gathering data. However, the accuracy of questionnaires can be affected by social desirability: the respondents' desire to create a positive impression of themselves.[159] Questions need to be derived with this in mind, and the answers are validated with other data collection methods such as observation, documentation, or focus groups.

DOI: 10.1201/9781003427681-6

Once the risks and vulnerabilities have been identified, mitigating recommendations can be created and prioritised to optimise available resources. The rest of this chapter contains recommendations to address people-related vulnerabilities.

ASSESS AND TRANSFORM ORGANISATIONAL CULTURE

Culture, as described in Chapter 5, refers to shared beliefs, attitudes, and perceptions about how to behave in an organisation. Understanding the organisation's cyber security culture is necessary to determine how to address people-related vulnerabilities. Reegård et al.[160] and Batteau[161] likened cyber security culture to safety culture. Lessons learned from the assessment and transformation of safety culture can be applied to enhance cyber security. Indeed, Nye[162] found that cultural norms against malicious cyber security behaviour could deter attackers.

Identify organisational culture types

Different sub-cultures within the organisation will have different effects on cyber security, so it is necessary to identify the different types. Various tools have been used to assess organisational culture, including the OCAI.[64] Culture types include bureaucratic, people-oriented, competitive, and innovative. Awareness of the vulnerabilities of each type (see Chapter 5) helps organisations tailor control strategies. For example, it may be necessary to review the number of procedures in a bureaucratic organisation and highlight the relative importance of cyber security ones. Provision of emotional support mechanisms, a characteristic of people-orientated cultures, could be beneficial to all organisations. However, people-orientated cultures would need to take particular care to prevent sharing of sensitive information and holding doors open. Competitive cultures, with an emphasis on productivity and winning business over cyber security, encourage workarounds. Therefore, procedural workarounds need to be anticipated, and a safer way provided does not inhibit the legitimate sharing of information. This means talking to employees and understanding their work requirements. Managers should allow sufficient time for employees to deal with emails and sensitive information. Innovative cultures may be more resistant to rigid cyber security procedures, so it is particularly important for them to involve employees in the design of the procedures.

Implement a transformation programme

Once the organisational culture type(s) has been identified, a transformation programme can be implemented. Several methods are available to support this process. In 2018, The National Protection Security Authority (NPSA, previously Centre for the Protection of National Infrastructure) published the "Secure 4"[163] questionnaires to assess security culture. Four surveys sought to identify culture type; employee perceptions of the organisation's cyber resilience; attitudes and need for education; and the frequency and causes of security behaviours. Ongoing development is underway at the time of writing.

Bremer,[164] proposed seven steps for organisational culture change. The first six steps seek to define the current culture and identify the *desired* culture and what needs to change. Employee engagement and raising awareness of the status quo and future vision are important parts of this process. The seventh step is where the plan to effect change is created. Participation from employees will help to create a sense of ownership and buy-in for this plan. Small groups of up to ten people were proposed. For the plan to succeed, Bremer advocates senior management commitment; leading by example; clear goals; employee commitment; continuous communication; correcting undesirable behaviours; and persistence.

Kotter and Cohen[165] created an eight-step model of culture change: establish a sense of urgency; establish a coalition of influential, lower-level managers; create a vision and strategy for change; communicate the vision and strategy repeatedly; remove obstacles; highlight progress but do not declare it too soon; maintain momentum; and facilitate improvements and constantly remind people of the benefits of the change. Lewin's change model (described in Burnes)[166] comprises three main steps. The first step, known as 'unfreezing', refers to changing the status quo in preparation for the desired change. Like Kotter, creating a sense of urgency is relevant here. The second step involves 'moving'; the implementation of the desired change. The third and final step is 'refreezing', stabilising the new norm or desired behaviours.

The UK Department for Environment, Food and Rural Affairs (DEFRA) produced the '4Es' (Enable, Encourage, Engage, and Exemplify) framework for behaviour change.[167] According to the model, first you need to enable change. This means removing barriers and providing any necessary facilities, such as storage for sensitive information, or investing in security software to facilitate change. Encouraging behaviour encompasses reward and recognition schemes. This could mean promoting the number and speed of cyber security incidents reported. "Engagement" refers to promotional media campaigns and employee participation. Exemplifying means leading by example and ensuring consistency in policies and procedures. Dolan et al.[168] added two more 'E's: "Explore" and "Evaluate", at the beginning and end of the change process, respectively. "Explore" refers to the initial activity to identify the behaviour to be changed. The "evaluate" stage is the analysis of the results of the change intervention.

All these models share the need to identify the desired change (e.g., specified cyber security behaviours) and communicate this vision. Employee involvement in shaping the change programme is another common factor. This allows people to highlight the impact of security procedure changes and design potential workarounds. Barriers or obstacles need to be removed and managers need to epitomise the desired behaviours. Paying 'lip-service' to cyber security while emphasising the need for greater productivity would be counterproductive. Instead, good behaviours should be recognised and incentivised.[158] Change should be communicated by respected peers.[169] The vision, progress, and benefits of the change need to be iterated, especially to avoid the change programme fizzling out over time.

Recommendations are summarised in Table 6.1. These findings correspond with motivation theories described in Chapter 3. For example, the recommendation for change to be communicated by respected peers has its roots in the influence of social norms. Involving employees in the change process relates to the impact of autonomy and intrinsic motivation. The removal of obstacles supports the identified need for coping efficacy to support motivation. Rewards and recognition were also highlighted as drivers of motivation.

TABLE 6.1 Organisational culture change recommendations

#	CULTURE TRANSFORMATION RECOMMENDATION	MORE INFORMATION
1	Identify and communicate inspiring vision and goals for change	For culture transformation to take place, the organisation first needs to understand its current and desired security cultures and the gap in between. The UK Government Cyber Security Breaches Survey 2022[170] found organisations were better able to implement cyber security changes when they were part of a wider programme to increase business resilience or efficiency. The smaller the change, the more robust the organisation during the transition. Adding additional changes in the first phase can cause more disruption.[171]

(Continued)

TABLE 6.1   (Continued)

#	CULTURE TRANSFORMATION RECOMMENDATION	MORE INFORMATION
		Objectives in support of the vision should have achievable relevant targets, be of interest to employees, supported by the organisation and rewarded. The need for the objectives should be explained to employees to allow them to accept and commit to them. Messaging needs to be clear and accessible.[172]
2	Respected managers and peers communicate the change	Advocated by Blanchard.[169] Coventry et al.[173] advocate the use of change champions and leaders to communicate. Maintain communications to keep momentum. Social media and other communication media can be used to positively influence the adoption of good cyber security behaviours.
3	Lead by example	Managers need to 'walk the talk', demonstrating the desired cyber security behaviours as opposed to simply paying 'lip-service'. McKenna[48], Bremer[164] and the board toolkit,[158] highlight the importance of support for the culture transformation from all levels of management.
4	Involve employees in the change process	Participative design of procedures and ways of working is likely to increase employee 'buy-in'. It also provides the opportunity to identify any potential obstacles and workarounds. Questionnaires and focus groups (of up to ten people)[164] can be used to this end. The 2022 Government Cyber Security Breaches Survey[170] found employees typically resisted change if they perceived it would make their job more difficult.
5	Remove barriers to change	This is necessary to enable change (see '4Es' model[167] described above).
6	Reward & recognise good cyber security behaviours	Recognising and publicising good performance or examples of the desired cyber security behaviours, allows people to feel good about themselves and creates a positive norm. According to motivation theory described in Chapter 3, the value of the reward for good cyber security behaviours should reflect the effort required to perform the behaviours. Rewards need to be allocated fairly and consistently and managers should keep their promises about behaviour rewards.

TABLE 6.1   (Continued)

#	CULTURE TRANSFORMATION RECOMMENDATION	MORE INFORMATION
7	Encourage blame-free incident reporting	A 'Just Culture',[51] allows employees to report when they have been the victim of a cyber security incident and when they have inadvertently disclosed information, without fear of blame or punishment. The NCSC board toolkit[158] also supports this. (See *Manage incidents*).
8	Analyse the impact of the change programme	It is necessary to identify what to measure from a potentially large pool of incident data.[174] Key Performance Indicators (KPIs) may facilitate reporting and communication. The NCSC board toolkit.[158] recommends that metrics should focus on success rather than failure. Therefore, measuring the number of employees who report phishing emails and suspicious behaviour, and how quickly they do this, is preferable to solely measuring the number of people who clicked on the malicious link. Employee engagement surveys can evaluate the impact of the change programme.
9	Reiterate vision, progress, and benefits of change over time	Identified successes from step 8 can be promoted to demonstrate the benefits of the programme and help prevent the change initiative from fizzling out.[175] Development areas can be addressed. Poor cyber security behaviours and workarounds need to be challenged so that they do not become norms. Organisational cyber security maturity can be measured and any further actions, determined (see *Conduct maturity assessment*).

DESIGN POLICY AND PROCEDURES AROUND USERS

As described in Chapter 5, if policies and procedures are not designed around the people who are expected to use them, they will find workarounds. Policy makers and procedure authors should therefore aim to make the most secure way to perform a task also the easiest way.

Investigate workarounds with users

Potential procedural workarounds should be identified by talking to users about their work and how the proposed procedures or security measures would affect them and their ability to perform their tasks. This may be achieved by focusing on groups and interviews. Care should be taken to avoid judgement and, instead, adopt empathy and understanding. Kirlappos et al.[176] reported that "heavy-handed" administration of policy by security personnel created tension among employees. A workaround investigation also provides the opportunity for security personnel to communicate why the precautions are necessary. Norcie et al.[177] found that ensuring whether users are aware of trade-offs between system performance and security was advisable. Once workarounds and their root causes are known, security personnel can identify a safer, controlled way of addressing the user's needs.

Encourage individual ownership of cyber security

In an ideal scenario, all employees would take responsibility for their own cyber security, rather than relying on technical security personnel and equipment. This message needs to be communicated to employees alongside education about cyber security. Allowing employees the autonomy to self-determine security behaviours should increase compliance with those behaviours and reduce workarounds.

Make policies accessible

The need to make cyber security policies accessible may sound obvious, but employees do not always know how to access their cyber security policy. One possible explanation is variation in terminology. Another reason might be that they simply do not know where to find it. There may be multiple policies referring to aspects pertinent to cyber security, such as social media policy, email policy, and data protection policy. This may confuse employees. The policies themselves may use technical language unfamiliar to non-security staff-members. Given what we know from motivation theory (see Chapter 3), the cost of performing good cyber security behaviours is likely to affect compliance, so onerous procedures are to be avoided. For this reason, it is necessary to simplify cyber security policies, use clear and consistent terminology, and make the policy easy for everyone to find.

Senior managers to endorse policy and procedures

Senior management endorsement of cyber security policy and procedures demonstrates that the organisation takes them seriously and therefore encourages employees to do the same. This is particularly important in competitive organisations, where productivity is recognised as a key driver. It is not enough to simply sign a name on the policy. It is also necessary for the managers to 'walk the talk' by exhibiting good cyber security behaviours and policy adherence.

CONSIDER HUMAN FACTORS IN RECRUITMENT AND STAFFING

Staff selection and allocation, job design, and reporting lines all factor in cyber resilience. Associated recommendations are given below.

Ensure the CISO reports to the board

If the Chief Information Security Officer (CISO) role is 'buried' within an IT department, senior decision-makers may lack visibility of cyber security issues and threats. It is advisable for them to report directly to the executive board. Otherwise, cyber threats and vulnerabilities may not be visible to senior decision-makers.[178,179] A lack of investment in competent security personnel will create vulnerability.

Employ enough competent security personnel

Inadequate investment in the recruitment and development of competent cyber security personnel is likely to result in burn-out and/or reduced effectiveness.[180] Sufficient numbers of competent security personnel need to be employed by the organisation. With the reported shortage of skilled security personnel,[5,10] development may be required. Workload issues should be mitigated by regular breaks from work and/or task rotation. Low workload can be as bad as high workload (Yerkes-Dodson in Cohen).[181] It can create vigilance effects and a loss of attention in security monitoring roles.

Screen candidates for roles with access to sensitive information

Security checks are common in organisations that require frequent access to sensitive information. They allow organisations to identify the potential for blackmail or coercion from malicious outsiders and should be conducted throughout employment. What is less common is routine personality screening to enhance cyber resilience. Advice regarding the ideal cyber security personality is provided in Chapter 4. However, as personality is complex and behaviour is situation-dependent, it should be used as a guide only. Personality testing is time-consuming, so for practical reasons, it may be necessary to focus the testing on high-risk personnel, such as those (candidates and extant personnel) with access to particularly sensitive information or those involved in cyber-attack response. Personality assessment can also be used to enhance understanding of different views and reactions and help tailor communications and training.

Design jobs to minimise access to sensitive information and ensure accountability

A good control mechanism is to prevent any one employee from accessing all the information necessary to commit cybercrime. For example, credit card companies may split customer information so that no one can access *all* the information pertaining to a customer's account. Another part of job design is accountability. Security teams identify fixes, but the IT development team implement them. Therefore, both teams need to be aware of who is accountable for which aspects of cyber security.

MANAGE THIRD PARTIES

It is important to ensure that associated third parties (suppliers, contractors, and customers) are aware of the organisation's cyber security policy and procedures, and that they take the necessary steps to protect themselves, and therefore, the organisation.[18] Commercial and procurement departments need to incorporate cyber security requirements and checks as part of contracting arrangements. Third parties need to be made aware of the organisation's cyber security policy and procedures to ensure the protection of hard copy, electronic and verbal data. Firewalls can be activated to restrict third-party access to information and critical systems.

ERGONOMIC DESIGN

Poor design of equipment and interfaces can give rise to errors. New technologies can enhance security and innovation, but they need to consider the human element in cyber security, by design.[182] This section outlines how to do that. The terms 'ergonomics' and 'human factors' are used interchangeably here.

Consider human factors in equipment design

Equipment that is used to monitor and analyse security, and indeed any application which comprises sensitive information, should be designed in accordance with human factors principles. Such principles have been developed since the 1940s and applied to enhance the efficiency and safety of systems. The need to consider the 'user experience' (UX) has gained popularity in recent years.[183] A traditional Human Factors Engineering (HFE) approach typically starts with some form of task analysis or user journey. The purpose is to break down the user's tasks to facilitate other aspects of design. A comprehensive task analysis should consider information requirements and outputs for each stage in the process, to inform decisions in all operating conditions (including abnormal situations such as a cyber-attack). The media for presenting this information can then be considered.

User workload capacity determines the amount of information one user can process and therefore the number of operators and displays required. Workload can be measured by various tools such as the NASA Task Load Index (TLX)[184] Bedford scale,[185] or a simple Likert scale. Decision aids can be provided to support complex information processing. The number of displays and operators determines the size and number of workstations needed. This, in turn, indicates the required size of rooms, such as Security Operations Centres (SOC), to accommodate the workstations and associated equipment. Human factors workplace design considers optimised team-working, visibility of any large-screen shared displays, access and egress requirements, and the working environment (lighting, temperature, noise, and vibration). To design a workstation, human factors engineers use anthropometric (body size) data pertaining to the target audience.[186,187] Common measurements include forward (or overhead) reach to controls; seated/standing eye-height (for viewing displays); buttock-to-popliteal length and hip/shoulder breadth (for seats); and fingertip diameter (for touchscreens). For practical reasons, human factors engineers typically exclude the upper and lower extremes of the population and often focus on the fifth to ninety-fifth, or the third to ninety-seventh, percentiles. The limiting dimension is adopted rather than the average. To illustrate this, imagine designing a set of shelves. The limiting measurement for access to the top shelf might be the fifth percentile female overhead reach, whereas the corresponding measurement for the lowest shelf would be the ninety-fifth percentile male lower reach, to prevent stooping. For a seated workstation, displays should be positioned to avoid a bent neck posture and excessive head and eye movement. Input devices or controls should be designed to consider anthropometry and biomechanics (essentially, the physics of how the body moves). The context of use is important. For example, touchscreen devices are difficult to operate while wearing bulky gloves.

Viewing angle limitations impact the positioning of displays.[149,188] The most frequently used displays/information should be positioned in the primary viewing cone. As well as protecting the user from musculoskeletal injury, this serves to maximise the likelihood that they will notice the required information. Where information needs to be displayed outside the primary viewing cone, blink coding and audible alarms can be used to attract attention. Viewing distance is combined with viewing angle limitations and eye-height, using trigonometry, to identify the optimal display area. Font and icon size should be driven by the viewing distance to ensure legibility.

Human Reliability Analysis (HRA) techniques such as the Human Error Assessment and Reduction Technique (HEART)[24] can be applied to areas identified as higher risk, to quantify error potential and associated mitigations.

It is advisable to test the arrangement of the workstations and their equipment with representative users in simulated operational scenarios. User-testing methodology is described later in this chapter.

Apply human factors to Human Computer Interface design

Security monitoring systems allow operators to detect unusual behaviour which might indicate an attack or attempt to compromise systems. This task involves processing large amounts of incident data.[174] Alarms design can have a big impact on the ability of users to observe and act upon suspicious behaviour and abnormal system events. Poorly designed alarms management systems can cause users (including security personnel,[104,189]) to miss important alarms because they are 'flooded' with information. In other words, there are too many alarms or pieces of information for them to cope with effectively. This can be avoided by prioritising alarms and diverting extraneous information away from the primary viewing area. Each alarm message should clearly indicate the operator's required action and be understandable, timely, and relevant to the recipient. Audible alarms should only be used to provoke an immediate response from the user, otherwise they become distracting. Operators have been even known to sabotage speakers to silence annoying audible alarms. This is understandable but could result in important alarms being missed. Audible alarm sounds should be distinct from other sounds or audible indicators in the vicinity to avoid confusion. Similarly, alarm text should be unique and distinctive. EEMUA 191[190] contains useful advice about alarm presentation.

Gestalt psychology tells us that people think in terms of 'wholes' to simplify the amount of information in their environment.[150] These pattern recognition skills are relevant for SOC personnel who monitor abnormal behaviour. The design of monitoring software needs to consider the potential for error in the interpretation of information, based on Gestalt principles such as proximity, similarity (perceiving close and similar items, respectively, as related), and closure (mentally forming shapes where elements are missing). Gutzwiller et al.[189] reported that SOC/security personnel lack feedback about the impact of their actions, making it difficult for them to understand how many attacks they have missed. Interface design should seek to enhance this feedback.

Colour-coding needs to consider conventions of the target audience. The use of red, amber, and green is a common user convention, with those familiar with traffic lights. However, these colours can be difficult for colour-impaired users to discern. This can be addressed by varying the depth and brightness of the colours and, importantly, using additional coding such as shape, hatching, and/or labelling to convey the meaning of the colours. Certain combinations, (such as bright red and blue) can cause unwanted visual effects and eyestrain, and others are difficult for colour-deficient users to distinguish,[191] so caution is advised. Different colours may mean different things to different users. Therefore, the choice of colours requires end user consultation as part of a participatory design process. Analysis of rail operators found nine colours were used to indicate alarm priority. In this case, it was decided to keep the colours when the system was upgraded, despite the cognitive demand required to process such a large number. This was because changing the colour-coding would likely result in error. Blink and auditory coding should be designed according to ergonomic principles. Their use should be limited to high priority information requiring a quick action, to minimise distraction.

Neilsen's ten usability heuristics[192] capture useful considerations for interface design including consistency, minimising memory load, error prevention and recovery, and provision of help.

Generative Artificial Intelligence (AI) has recently been used to write software code and support the design of products. It can also constitute a threat, if employees disclose sensitive or proprietary information in an open the platform. The design process therefore needs to consider the potential risks. Gutzwiller et al.[189] cited automation bias as a risk with the use of more automated, professional cyber security technology such as AI.

Conduct user-testing

A participatory design process requires proper scrutiny of designs by representative users. Access to users can be difficult. The temptation might be to use people with previous, possibly out of date, experience of a product. This tactic is preferable to the absence of consultation. However, it is no substitute for proper user representation. The discipline of human factors was built on the premise that equipment or product designers may not be the best people to test their own equipment. They may be expert users or extremely familiar with the product, which gives them an advantage over typical users and can yield inaccurate test results.

Usability of software and hardware equipment can be assessed in terms of effectiveness and efficiency. Effectiveness can be defined as accuracy determined by measures such as the number of errors and tasks completed correctly. Efficiency is defined as the measure of effectiveness divided by the resources used in achieving the level of effectiveness.[193] Example measures include time taken to perform a task, cost, and fatigue. It may not be possible to achieve the perfect design. There may need to be trade-offs between system performance and security. While there is no definitive pass mark for usability, several tools are available to measure it. They are normally applied after or during the performance of a representative scenario(s) using the equipment under assessment. The System Usability Scale (SUS)[194] is questionnaire containing ten items, including ease of use and intuitiveness. Users score each item on a Likert scale, from 'strongly agree' to 'strongly disagree'. The Situational Awareness Rating Tool (SART),[195] also incorporates a ratings scale. It considers factors such as the stability, complexity, and variability of the situation, as well as attentional aspects, alertness, and mental capacity. Less formal methods such as open-ended questioning and observation can also derive valuable results. As well as usability measures, workload assessment tools, mentioned earlier, can be applied as part of user-testing, to determine mental and physical burden when using equipment.

Consider human factors in the procurement and selection of equipment

Budget and time constraints mean it is not always possible to consider human factors in the design of equipment. However, the selection of Commercial Off The Shelf equipment (COTS) also needs to consider human factors requirements. Checklists (designed by competent human factors personnel) can be provided to suppliers to identify the extent to which relevant considerations, such as user consultation, have been adopted. As mentioned in Chapter 4, human factors research needs to keep pace with technological developments to mitigate pro-innovation bias.

APPLY HUMAN FACTORS PRINCIPLES TO REPORT PRESENTATION

The choice of colours is an important consideration in the design of reports, as well as HCI. Cyber security information reports may be electronic or paper-based. In both cases, their design needs to consider the target audience. Security personnel will need more detailed information than regular board members. Busy senior managers may need to simplify cyber security information, along with all the other important information they process. Therefore, the information needs to be tailored to suit different needs, using non-technical language for board executives, to help them understand the risks and implications of their decisions. If reports are too complex, they may be ignored.[137] Language also requires consideration as it has been shown to affect behaviour.[123] Information that is presented to align with the overall business strategy is more likely to engage the board.[158]

MANAGE ASSETS

Incidents such as 'WannaCry', a ransomware attack in May 2017,[196] highlight the need to invest in an up-to-date operating system, regular software patching, and antivirus protection. Affected organisations included the National Health Service (NHS) in the UK, which was forced to drastically reduce services. Asset management systems need to accurately record up-to-date information about the status of software and hardware equipment. Redundant accounts need to be routinely closed to minimise the risk of unauthorised access, especially from disgruntled leavers.

DISCOURAGE USE OF PEDS FOR WORK

No matter how well an organisation implements controls to protect work equipment, it is not possible to control information shared on Personal Electronic Devices (PED), as described in Chapter 5. Therefore, the use of PEDs for work should be discouraged.[6] Investment in usable, encrypted, portable work devices may reduce the use of personal equipment for business activities. Encryption decreases the risk of information loss from a misplaced laptop.

MANAGE PASSWORDS, EMAIL, AND THE INTERNET

Passwords as a means of user validation; misuse of email; and the internet, are sources of vulnerability.

Allow users to choose their passwords

Mandating complex, lengthy passwords that change frequently can be counterproductive. Giving users the autonomy to choose their own passwords may increase the likelihood that they will remember them.[54] Password safes, allowing users to store multiple passwords, should be encouraged.

Investigate alternatives to passwords

Passwords create vulnerability so it is advisable to investigate alternatives. Biometric technologies, such as facial or fingerprint recognition may be a better solution because they eliminate reliance on human memory. Fingerprint and facial recognition systems are common on smart phones but could be better utilised in commercial computer equipment and applications. However, resistance may be encountered where users are concerned about their privacy and personal data protection. So, it may be advisable to use biomechanics in conjunction with other methods of authentication (Corrella and Lewison [in Moallem]).[91]

Behavioural metrics such as keyboard stroke dynamics, which analyses the timing and rhythm of a person's typing, may also support user validation.

Manage email

Phishing emails may be missed because of the volume of emails received. While training and phishing simulations may help, the root cause of the issue may be addressed by reducing the number of emails received, by investigating alternative means of communication, and encouraging greater time allocation for managing them. Better still, technological solutions should detect malicious emails and prevent them reaching the target(s) in the first place. Segmentation of information permissions across individuals and/ or departments minimises the impact of access to one account. A Virtual Private Network (VPN) is advocated by the NCSC[197] to protect data sent and received, through real-time encryption. Emails and USB usage can be monitored, but the root cause of the problem should be investigated.

Manage Internet usage

High-risk websites should be automatically blocked on work devices. Software downloads can be managed by restricted administrator rights to reduce malware risk. Work-related information shared in online social media forums and generative AI sites can be monitored by Open-Source Intelligence Surveys (OSINT).[198] A generative AI policy should be created and communicated to employees.

Manage tele/videoconferencing

Restrict the type of information that can be shared on specific tele/videoconferencing platforms and ensure employees are aware of the risks associated with sharing information on each.

PROVIDE CYBER SECURITY TRAINING AND AWARENESS

When employers ask why people 'fall for' phishing emails, clicking on malicious links or attachments, and even submitting personal information or login credentials, training is often posed as a solution. The other recommendations in this chapter demonstrate that it is not the only one. However, it is indeed important to equip employees with the knowledge and tools to support them against attacks and to ensure they feel confident in their ability to protect themselves and the organisation. All departments within an organisation need to be aware of their impact on cyber security.

Capture motivation

"Human error" continues to be cited as a cause of many cyber security incidents despite training, so something is going wrong. To avoid people simply clicking through material without fully absorbing it, the training needs to capture the intrinsic motivation of the trainee,[173] such as learning, accomplishment or sensory pleasure.[55] Including behaviours they can adopt at home, to protect their family's personal devices, gives them another reason to pay attention. Telling people not to give their information to people they do not trust is stating the obvious. They need to understand the *importance* of good cyber security behaviours. A good culture should help. The aim of training is to frighten but provide solutions to make trainees feel capable of protecting themselves from attacks in a cost-effective way.

Frequent attacks can be demotivating, so training should seek to create positive associations with good behaviours by recognising them. This may be particularly important for security personnel who deflect one attack, only to be faced with more to deal with, in a never-ending cycle.[189] Training should deter overreliance on others, such as the IS department, and instead, encourage people to take ownership of cyber security, for example, through participatory design.

See the TOMS model in Chapter 3 for a summary of motivation theories. Individual differences, such as personality (as discussed in Chapter 4) and how each type can be affected, should be considered.

Incorporate realistic examples

To avoid the human tendency to downplay attacks, believing them to happen to other people (optimism bias), the training needs to incorporate believable examples that are relevant to the target audience. The examples should not be alarmist, because people may become demotivated or (if nothing bad happens) desensitised to warnings. The chances of occurrence should also be conveyed. Employees should be made aware of the frequency and severity of cyber-attacks on their organisation. Awareness initiatives should convey the impact of attacks on the organisation and employees themselves.[199]

Examples should be tailored to the target audience, bearing in mind that there may be several different user types within an organisation. For example, a Human Resources (HR) department could be given examples of the consequences of fake CV attachments or mismanagement of personal data, whereas the marketing department could be alerted to the pitfalls of sharing sensitive information.

Attack examples need to be refreshed and recent, to be easily recalled when a potential threat is encountered. Long-term consequences such as reputational damage should be highlighted as well as immediately obvious effects.

Make training engaging

One of the reasons people ignore training is because it is not engaging enough. Gamification has been successfully used to engage trainees. SoSafe[17] reported that it improved training activation by "up to 50 percent". They later advocated regular nudges (gentle reminders about behaviour), combined with gamification, to help engagement.[200] Ideally, employees will eventually perform the desired behaviours out of habit.

Techniques observed in gaming such as scaffolding (progressive disclosure) can be used to move trainees towards greater levels of knowledge and skills. Adaptive training responds to the user's needs.[189]

Training could take the form of personnel at entry points wearing a badge or clothing prominently depicting the word "intruder" or the like. This could be humorous and memorable. It sends a message that tailgating is not acceptable and challenging tailgaters and reporting intruders is encouraged.

Types of training

Some specific types of training and education are categorised below.

Provide incident response training

Relevant personnel, including stakeholders from IT, legal, marketing, and senior management, need to understand the incident response plan. Role-play and table-top exercises allow the response leadership team to practise their reactions and understand each other's behaviours. The training needs to alert trainees to the risks of different courses of action and the impact of poor situational awareness - dealing with the immediate threat and not the bigger picture; making decisions with insufficient information; adrenaline-fuelled impulsivity; and risky decision-making, such as the potential human tendency to avoid shutting down internet systems despite a greater risk of loss, as discussed in Chapter 4.

Trainees should understand the role and perspectives of all the stakeholders and the importance of allowing everyone to voice their concerns. The IT team may wish to shut the internet down, whereas the operations leaders may be reluctant to do this. Personality differences can also play a part. A deliberator may want more information before making a decision than a fast-decision-maker would.

Conduct phishing training

Microsoft, 2023, provided advice to detect malicious phishing emails.[201] Suspicious characteristics cited include a sense of urgency; unfamiliar sender; poor spelling and grammar; lack of a personalised addressee name (instead using "Sir", "Madam" or "Hello", for example); and an incorrect sender email address. They advised that links in emails can be checked by hovering the mouse over them instead of clicking them. Lallie et al.,[202] found malicious phishing emails were more successful during the COVID-19 pandemic if they used media and government announcements. This created a sense of realism and authority and can make fakes hard to detect. Verizon[13] proposed measures of effectiveness for training in phishing email detection, including incorporating a measurable outcome, such as the number of correct answers on a questionnaire after the training.

Internal phishing campaigns are used to measure the number of employees who click, provide information, and/or report the fake suspicious email. However, too many phishing simulations can cause resentment, and desensitise employees so they miss a genuine attack. Private Eye[203] reported the anger of News UK employees after a phishing simulation. They should therefore only be used in moderation and in conjunction with other solutions.

Provide training in AI risks

AI technology brings benefits such as analysis of threat data and trends. Generative AI tools like "CHATGPT" can be usefully exploited to write code and identify programming errors. However, this risks the loss of an organisation's intellectual property and passwords. AI tools like *"WormGPT"* can be used to create phishing emails to rapidly distribute malware. "Deepfakes" mimic images, videos, and voices of influential people to facilitate social engineering attacks. AI can also be used to spread fake news and propaganda. Awareness training needs to incorporate these risks.

Provide data protection legislation training

Training in data protection legislation, such as the UK Data Protection Act 2018 and GDPR,[2] and associated internal policy, should reduce the risk of inadvertent loss of sensitive information. The training could include warnings about how to respond to 'pop-up' messages, including those pertaining to privacy and personal information.

Provide training in physical environment security

Training to address physical environment vulnerabilities, discussed in Chapter 5, can include visitor management awareness; attacker tactics like holding coffee cups or walking on crutches to entice people to hold secure doors open for them; the need to phone security directly rather than a potentially false number on a letter of accreditation; and warnings about discussing work in public spaces. Employees can be trained to show their identity pass/ credentials if someone holds a door open for them, to avoid looking suspicious. This takes pressure, to request credentials, off the person in front. The dangers of free public Wi-Fi (wireless fidelity) should be explained.

Provide training in password management

Employees need to understand the organisation's password policy and the need for strong passwords (Adams and Sasse 1999 in Ertan et al.).[199] They can be advised to use a password safe to reduce the need to remember multiple complex passwords. Alternatives or additions to passwords such as biometrics can be encouraged.

Provide training in social engineering tactics

Employees need to be made aware of techniques employed by attackers designed to persuade them to part with sensitive information or valuable assets. Coatesworth,[204] cited seven principles of influence identified by Cialdini (1984) and adopted by social engineering attackers. They rely on people's tendency to repay a debt or gift (reciprocity); maintain a course of action (commitment and consistency); follow others' behaviour (consensus); follow experts and leaders (authority); agree to those we like (liking); value things that are in short supply (scarcity); agree to those we identify as similar to ourselves (unity); and succumb to flattery. The dangers associated with pop-up messages can be included as people have been found to spend very little time reading them.[101]

Allocate time for training

A good reason for failure to complete or pay attention to cyber security training is the lack of time available to do so. Managers need to be supportive of training and allow employees adequate time in which to complete it.[205] If training is treated as a task to be accomplished in the employees' own time, separate from core business, it may be neglected. Instead, employees can be encouraged to block out sections of their working time to dedicate to training and treat it as a priority activity. This may require a culture shift and senior manager endorsement. To complete with other training and work demands, the cyber security training should be flexible in design, and fit around busy schedules.

Manage frequency and timing of training

Cyber security training is advisable during recruitment/induction. However, if it is not repeated, employees may suffer from skill-fade, and forget aspects of the training over time. Threats may change over time too. Therefore, refresher training needs to be conducted throughout employment. Regular prompts to complete small chunks of training material, with tests, may be more practical than lengthy training courses. A Training Needs Analysis (TNA),[206] can determine the appropriate frequency of training, based on how often risky tasks are performed and the level of task and/or threat complexity. Outside employment, people need training in the dangers likely to be encountered online. Witsenboer et al.[207] recommended that formal cyber security education should begin at school.

Create and maintain a competence management system

A competence management system gives organisations awareness of how many employees have undertaken cyber security training, when and to what level of understanding. Training should incorporate a test to measure competence and understanding in the material they have seen. A pass mark could be included to allow employees to retake the test until they achieve an acceptable score. Multiple choice questions facilitate scoring but, of course, if they are too easy, trainees can achieve the correct result by trial and error. The ECCC, 2023 Strategic Agenda[182] includes an action for a cyber security skills framework and competence assessment.

SECURE THE PHYSICAL WORKING ENVIRONMENT

As discussed in Chapter 5, the physical working environment can be a cyber security vulnerability. Biçakci and Evern[208] advocated communication and a joint reporting between the physical, and cyber, security departments, in the context of nuclear power plants. Traditionally Operating Technology (OT), such as those found in those plants, was protected by an 'air-gapped' system which could only be breached by physical access by malicious insiders or infected USBs. It is not only OT that is at risk from physical environment vulnerabilities, however. Any system or organisation is at risk of loss of information through this route. Further recommendations are below.

Secure entry points and server rooms

Physical penetration testing can be used to test the security of a building. An independent actor poses as an unauthorised attacker attempting to gain entry and remove information without access credentials. They may hold a letter of approval but it can contain a false phone number to see whether the challenger will use their directory to access the correct number/contact. Otherwise, if they call the false number, another actor can pretend to verify the person in front of them. The results of the test are discussed with the employer, along with recommendations to enhance their physical security.

Social compliance, politeness, and even laziness can inhibit entry-point security. Turnstiles may be an expensive solution, but they do not rely on employees to challenge anyone who might be attempting to gain unauthorised entry. Intermittent presence of visible security personnel at access points provides a clear allocation of responsibility for checking credentials and allows other employees to transfer this responsibility away from themselves (see Latené and Nida).[34] Notices on doors reminding employees about tailgating reinforce this requirement and may help alleviate some of the social embarrassment surrounding tailgating prevention.

Locked server rooms provide an extra barrier after outer entry points. Further layers of security, derived from practical experience, help protect the organisation if a malicious outsider manages to gain access to a building or sensitive area, as follows.

Lock computers

Encourage employees to lock their computers when they leave their desks, to prevent network access. Provide screen privacy filters for monitors to protect information from 'shoulder surfing': being observed by others.

Provide lockable storage for sensitive paperwork

To enable safe storage of sensitive hard copy information at home and in the office, lockable storage facilities should be provided. Measures should be taken to protect hard copy information when in transit.

Prevent remote printing

To avert unauthorised access to hard copy information left on printers, avoid the ability to print-and-forget. This can be achieved by requiring employees to input a password or Personal Identification Number (PIN) on the physical printer before collecting their paperwork. This process should not be onerous, otherwise it may encourage workarounds.

Dispose of sensitive documents

For confidential or sensitive hard copy material, provide a special means of disposal such as a clearly marked bin, and ensure the material is properly shredded or otherwise destroyed.

Practice and monitor good housekeeping

A clear-desk policy should be implemented to reduce the risk of unauthorised access to paperwork. Employees should be reminded to lock their computers when they leave their desks to protect electronic information and networks. At night, blinds should be closed and unnecessary lights, switched off in ground floor (and other visible) areas, to prevent visual access to information through windows. Regular housekeeping inspections should monitor information left on desks and printers, whiteboards, unlocked computers, lights, and blinds.

Provide secure spaces for sensitive conversations

Where possible, create a separate area to protect sensitive verbal discussions. This includes information transmitted via videoconferencing, or calls when working at home or other remote locations such as trains, planes, buses, and cafes. Where this is not practical, take steps to minimise the risk of information being overheard by unintended personnel, and avoid mentioning the company name in these circumstances.

Discourage wearing of company lanyards and identifiable clothing in public

Wearing company-identifiable clothing or lanyards creates vulnerability. Sometimes it will be necessary, for example at a promotional event. In these circumstances, staff should be advised to be careful about the information they discuss.

Manage visitors

If employees are used to seeing people that they don't know in their working environment, it makes it difficult for them to notice an unauthorised attacker on the premises. Therefore, visitor management policy should make obvious who is a genuine visitor and whether they need to be escorted by a member of staff. This can be achieved by identifiable badges. Booking visitors in advance and requiring identification is also advisable.

MANAGE INCIDENTS

Cyber security attacks happen all the time with varying degrees of success, depending on the resilience of the targets. Incident analysis should consider near misses as well as more damaging attacks. Given the large proportion of incidents attributed to human error, consideration of human factors causes and contributary factors is likely to enrich investigations and enhance resilience. Organisations need to be able to monitor, anticipate, respond to, and learn from incidents.[209] Figure 6.1 illustrates the proposed incident management cycle. When incidents have been identified, the defensive response takes place. After the incident, an investigation facilitates learning which can be used to identify further incident potential.

The following paragraphs expand on these stages.

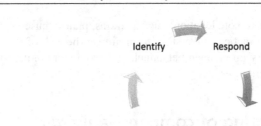

FIGURE 6.1   Incident management cycle

Identify

Share incidents

To identify potential incidents, it is necessary to share attack trends. Organisations should monitor threat intelligence and be aware of how often they are being attacked and the level of success of each attack. Likewise, employees may not realise the scale, likelihood, and consequences of an attack, so employers need to share this information with them as part of awareness and education. Incident reports should be communicated to the board and employees in a way that they understand and can action, as described earlier under *report presentation*.

Facilitate incident reporting

An organisation cannot manage and prevent incidents if they do not know that they are happening. Therefore, the first step is to encourage people to report incidents without punishment and in a timely manner. Incident reporting needs to be easy to access and quick. For example, a button in the email application could provide direct access to report phishing emails. A link to report other security concerns or attacks could be placed in a prominent location on a company intranet front page. A confidential incident reporting system may encourage employees to come forward with concerns about cyber-attacks or their colleagues' behaviour. They may be more likely to share incident information if they believe it is useful and the sharing will be reciprocated.[210]

Measure incidents

An incident database can be created to monitor internal and external threats and vulnerabilities. Some example incidents are described in Chapter 2. Internal incident metrics such as the number of people who report incidents, and the speed with which they do so, are preferable to the number of people who engage with a malicious phishing email and/or fail to report it (NCSC board toolkit).[158]

Identify the number of lost/stolen laptops and other devices

Lost or stolen company equipment needs to be recorded so the scale of the risk of data loss can be known.

Share and monitor incidents externally

To enable organisations to anticipate attacks, they need to be able to learn from others.[182] Unfortunately, it can be hard to obtain root causes of incidents where companies are reluctant to share them. It is understandable that businesses may not wish to expose their vulnerabilities and open themselves to reputational damage. However, this information is vital for effective threat intelligence. The NIS2 directive (2023)[211] requires companies within the European Union (EU), or operating with EU organisations, to share information to enhance cyber security. In the USA, the Cybersecurity & Infrastructure Security Agency (CISA) encourages information sharing.

Respond to incidents

Organisations need to know how to respond during an attack. Senior managers need to know what information to obtain when, and from who. Having a CIO on the board should help. Managers need to be aware of the potential human tendency to avoid shutting down internet systems despite a greater risk of loss, as discussed in Chapter 4. Cyber security personnel may favour a different course of action to operations leaders who are concerned with productivity. Response team leaders therefore need to listen to the opinions of all members to avoid a decision that is led by a few dominate characters. To facilitate an efficient incident response, a plan and procedures can be devised in advance. Role-play training can be a valuable tool to test and hone incident response skills, as discussed earlier.

Investigate incidents

Once an incident has happened, an incident investigation should identify why it did, and devise recommendations to reduce the risk of its reoccurrence. Where organisations receive many attacks, it may be necessary to prioritise investigations to focus on common susceptibilities. Root cause analysis methodology can be applied to identify common causes of incidents. This approach is best achieved with a multidisciplinary team, including competent human factors specialists, as well as cyber/information security, IT, and operations representatives. There are a number of models that can be used to inform root cause analysis such as Toyoda's 'Five-Whys' technique.[212] This concept, of repeatedly asking why each step in an incident happened, is adopted by Ramsey's, Kelvin TOP-SET[213] incident investigation approach, which categorises incident condition indicators, in terms of Technology, Organisation, People, Similar events, Environment and Time sequences, to provoke thought and understand the incident architecture. Attacker personas can be combined with threat and risk modelling to present architectural views of an incident.[214] Another incident architecture framework is Stanton's.[215] This has been applied to identify where links between darknet tasks, social agents, and information could be broken, to disrupt buying and selling activities.[216]

Experience in incident investigations has unfortunately observed that training has been recommended as a solution for so-called "human error". It is important to remember that training is not the only solution. It can be patronising to mandate training for someone who is experienced in their job, and it does not get to the root of the problem – the reasons for their behaviour. For example, a person may click on a malicious link in an email even when they know links can be dangerous. This could be because it was convincing and they thought it was genuine; they were in a hurry and did not give the email their full attention; or receiving genuine links in emails was commonplace in their work environment. Phishing emails can be difficult to spot and technological advances may exacerbate the problem. A study by Lim et al.[15] reported that emails written by AI were more successful than those written by humans. In another example, an employee may deliberately breach email policy because they are under pressure to share a deliverable report in a hurry and the security measures prevent this. Providing an employee with training in a policy that does not support their work may not, therefore, prevent recurrence. Instead, investigations should consider the organisational root causes and human vulnerabilities such as those described earlier in this book, including analysis of mental short-cuts, cognitive biases, and known error-producing conditions. This should allow investors to generate more impactful recommendations to reduce the likelihood of future incidents. The findings of the investigation need to be shared with employees and other organisations, to raise awareness of the risk and enable mitigation.

Learn

Organisational learning relies on the review and implementation of lessons from incidents.[217] It is too easy to file an incident report and forget about it. Investigations should generate Specific, Measurable, Achievable, Specific, Relevant, Time-bound (SMART)[218] actions. Actions might include updating policies, procedures, and training material; system development; and a resource review to ensure there is sufficient access to expert knowledge. These actions should then be implemented, and relevant reminders reviewed at the onset of a project and when setting up systems. Employees should be encouraged to challenge processes and procedures based on their experience, thereby creating a double-loop learning environment[219] and facilitating continuous improvement. Learning teams can be established to share operational intelligence between the people who do the work and the people who design it. This enables understanding of the actual work done instead of assuming people simply follow the procedures.

MANAGE MALICIOUS BEHAVIOUR

As discussed in Chapter 5, reasons for employees to behave with malicious intent include lack of perceived fair reward, job security concerns, financial or emotional difficulties, desire for more excitement, and disagreement with organisational principles. It is therefore advisable to monitor triggers such as these. Employee engagement surveys can indicate the general mood. Managers and HR personnel can inform the security team about high-risk indicators including employment terminations; frequent changes of position; unpopular policies; and avoidance of vetting and screening. Other indicators to monitor include logon and swipe access at unusual hours of the day/night; large spends and/or late payment of company credit card bills; and external email communications.

Emotional support mechanisms should be provided for employees. The responsibility for the duty of care of individuals needs to be clear, especially in a matrix management organisation. It should not simply be seen as an HR issue.

CONDUCT DECEPTION

The term "deception" is used in cyber security to describe the use of "honeypots" to divert an attacker's attention away from genuine assets. A honeypot is a fake asset, designed to look like another machine on the network. When it is attacked, the organisation is alerted. Fake user accounts and credentials can be incorporated into systems to entice attackers. They can be named to suggest high value targets such as administrator accounts. Security personnel can identify when fake credentials have been used to attack an asset. AI, in particular Machine Learning, can be used to detect anomalies that are indicative of an attack.[220] Of course, attackers know about honeypots. They just need to decide how much risk to take to access the genuine assets.[221] A study by Kranenbarg et al.[94] indicated that cybercriminals were conscientious, diligent, and brave. This suggests they may persist with an attack despite the risks. As discussed in Chapter 4, people are less likely to take risks if they expect a gain as opposed to a loss. The perceived gain may depend on the attacker's motivation. The use of psychologists/human factors professionals in deception design is recommended.

CONSIDER HF IN OFFENSIVE CYBER SECURITY

Smeets[222] defines offensive cyber security as activities to "disrupt, deny, degrade, and/or destroy" a target computer-based infrastructure. It is not just the domain of cybercriminals. It may also be employed by state actors as part of intelligence activities. A common attack model comprises four stages: penetrate, control, embed, and connect.[223] Human motivation, biases, and vulnerabilities (individual and organisational) described in this book may enhance an offensive cyber security strategy as well as a defensive one.

CONDUCT MATURITY ASSESSMENT

The assessment materials in Chapter 7 can be used to establish the current organisational cyber security situation and a measure of maturity. Then, the recommendations in this chapter can be interpreted to determine mechanisms to move the organisation towards the desired state. A transformation programme can drive this. After the relevant recommendations have been implemented, maturity can be measured again, to hopefully demonstrate a reduction in risk and increase in maturity. The CIEHF's HACs framework[224] mapped human factors to the NIST[225] maturity framework. A new framework has been developed to measure human-related cyber security maturity: GRADE. It comprises five levels: Growing (1), Reactive (2), Alert (3), Developed (4), and Enhanced (5). An illustrative example is provided in Chapter 7 (Table 7.8). In the lower levels, most employees lack awareness of the need for cyber security, are not consulted about procedures, and may be blamed for security incidents. Higher maturity is associated with enhanced employee understanding and ownership of cyber security, a range of performance indicators, and continuous improvement.

NOTES

156 Institute of Internal Auditors, 2020, *The IIA's three lines model, an update of the three lines of defense.* Institute of Internal Auditors (pubs).

157 NCSC, 2021, *10 Steps to cyber security*, Version 1, https://www.ncsc.gov.uk/collection/10-steps. National Cyber Security Centre, UK (pubs).

158 National Cyber Security Centre, 2023, *Cyber Security Toolkit for Boards*, Resources designed to help board members govern cyber risk more effectively.

159 Van de Mortel, T.F., 2008, Faking it: Social desirability response bias in self-report research, *Australian Journal of Advanced Nursing* 25(4), 40–48.

160 Reegård, K. and Blackett, C., 2019, The Concept of cyber security culture, *Proceedings of the 29th European Safety and Reliability Conference*, Beer, M. and Zio, E. (Ed.s), Hannover, Germany.

161 Batteau, A.W., 2011, Creating a culture of cyber security, *International Journal of Business Anthropology* 2(2), Sun Yat-Sen University (pubs). https://doi.org/10.33423/ijba.v2i2.1179

162 Nye, J.S., 2017, Deterrence and dissuasion in cyberspace. *International Security* 41(3), 44–71. https://doi.org/10.1162/ISEC_a_00266.

163 NPSA, 2018, *Introduction to SeCure 4*, https://www.npsa.gov.uk/secure-4-assessing-security-culture (accessed 2023).

164 Bremer, M., 2012, *Organisational culture change, unleash your organization's potential in circles of 10* (1st ed.). Netherlands: Kikker Groep. ISBN: 978-90-819825-1-1.

165 Kotter, J.P., & Cohen, D.S., 2002, in McKenna, E., 2006, *Business psychology and organisational behaviour* (4th ed.). Psychology Press. ISBN 10: 1-84169-392-8.

166 Burnes, B. 2004, Kurt Lewin and the planned approach to change: A re-appraisal, *Journal of Management Studies* 41(6), 0022–2380.

167 Defra, A. 2008, Framework for pro-environmental behaviours, Report: https://assets.publishing.service.gov.uk/media/5a789f08ed915d04220640a4/pb13574-behaviours-report-080110.pdf.

168 Dolan, P., Hallsworth, M., Halpern, D., King, D., Vlaev, I., Cabinet Office, Institute for Government, Mindspace. *Influencing behaviour through public policy, the practical guide*, https://www.instituteforgovernment.org.uk/sites/default/files/publications/MINDSPACE.pdf.

169 Blanchard, K., Britt, J., Zigarmi, P. and Hoekstra, J., 2008, *Who killed change? Solving the mystery of leading people through change.* Harper Collins. ISBN 978-0-00-731749-3.

170 Department for Digital, Culture, Media & Sport, 2022, Official Statistics: Cyber Security Breaches Survey 2022, https://www.gov.uk/government/statistics/cyber-security-breaches-survey-2022/cyber-security-breaches-survey-2022.

171 Hingst, L., Ast, J. and Nyhuis, P., 2022, Framework for assessing the impact of change on a factory by adapting learning behaviour models, *Procedia CIRP 107 (2022) 393–398, 55th CIRP Conference on Manufacturing Systems*, Elsevier.

172 Krusche, A., Wilde, L., Ghio, D., Morrissey, C., Froom, A. and Chick, D., 2022, Developing public transport messaging to provide crowding information during COVID-19: Application of the COM-B model and behaviour change wheel, *Transportation Research Interdisciplinary Perspectives* 13(2022), 100554.

173 Coventry, L., Briggs, P., Blythe, J. and Tran, M., 2014, Using behavioural insights to improve the public's use of cyber security best practices, Summary report, Government Office for Science, URN GS/14/835.

174 Henshel, D., Cains, M.G., Hoffman, B. and Kelley, T., 2015, Trust as a human factor in holistic cyber security risk assessment, *Procedia Manufacturing* 3, 1117–1124.

175 Kotter J.P. and Cohen, D.S., 2002, in McKenna, E., 2006, *Business psychology and organisational behaviour* (4th ed.), New York: Psychology Press. ISBN 10: 1-84169-392-8.

176 Kirlappos, I., Parkin, S. and Sasse, M.A., 2014, Learning from "Shadow Security": Why understanding non-compliant behaviours provides the basis for effective security, *Workshop on Usable*.

177 Norcie, G., Blythe, J., Caine, K. and Camp, L.J., 2014, Why Johnny can't blow the whistle: Identifying and reducing usability issues in anonymity systems, *Workshop on Usable Security*, https://doi.org/10.14722/usec.2014.23022.

178 Carder, J., 2021, Why your CISO should report directly to the CEO, *Forbes*, https://www.forbes.com/sites/forbestechcouncil/2021/09/08/why-your-ciso-should-report-directly-to-the-ceo/ (accessed 2024).

179 Lowry, M.R., Sahin, Z. and Vance, A., 2022, Taking a seat at the table: The quest for CISO legitimacy, *ICIS 2022 Proceedings*, 14, https://aisel.aisnet.org/icis2022/security/security/14.

180 Gutierrez, L., 2021, Why managing the human factors is crucial to a successful cyber security crisis response, *PWC*, https://pwc.blogs.com/cyber_security_updates/2021/03/index.html.

181 Cohen, R.A., 2011, Yerkes–Dodson Law. In J.S. Kreutzer, J. DeLuca, B. Caplan (Eds.), *Encyclopedia of clinical neuropsychology*. New York: Springer. https://doi.org/10.1007/978-0-387-79948-3_1340.

182 European Cyber security Competence Centre (ECCC), 2023, Strategic Agenda.

183 Luther, L., Tiberius, V. and Brem, A., 2020, User Experience (UX) in business management, and psychology: A bibliometric mapping of the current state of research, *Multimodal Technolologies Interaction* 4(2), 18, https://doi.org/10.3390/mti4020018.

184 Human Performance Research Group, National Aeronautics and Space Administration (NASA), 1980s, Task Load Index (TLX) Version 1.0 Paper and Pencil Package. https://humansystems.arc.nasa.gov/groups/tlx/index.php.

185 Casner, S.M. and Gore, B.F., 2010, *Measuring and evaluating workload: A Primer*, NASA/TM-2010-216395.

186 BSi British Standards, 2010, ISO/TR 7250: *Basic human body measurements for technological design, Part 2: Statistical summaries of body measurements from ISO populations.*

187 Cummings, R., 2022, *Human factors integration technical guide for anthropometry: Peoplesize*, Version 4.4.

188 ISO, 2013, BS EN ISO 11064-4: Ergonomic design of control centres, Part 4: Layout and dimensions of workstations Edition 2, Technical Committee : ISO/TC 159/SC 4, BSi (pubs).

189 Gutzwiller, R.S., Fugate S., Sawyer, B.D. and Hancock, P.A., 2015, The human factors of cyber network defense, In *Proceedings of the Human Factors and Ergonomics Society 59th Annual Meeting - 2015*, LA, CA.

190 EEMUA, 2013, Engineering Equipment and Materials Users Association *(EEMUA) Publication 191 Alarm systems – a guide to design, management and procurement*, ISBN 9780859311922.

191 BS EN ISO9241-112 BSI, Ergonomics of Human-System Interaction (2017) Part 112; Presentation of Information.

192 Nielsen, J., 1994, Enhancing the explanatory power of usability heuristics. *Proceedings ACM CHI'94 Conference.* Boston, MA, April 24–28, 152–158.

193 ISO 9241-11: Usability: Definitions and Concepts and 5 ISO/IEC 25022: Measurement of Quality in Use.

194 Brooke, J., 1995, *A quick and dirty usability scale*, https://www.usability.gov/how-to-and-tools/methods/system-usability-scale.html.

195 Selcon, S.J. and Taylor, R.M., 1989, Evaluation of the Situational Awareness Rating Technique (SART) as a tool for aircrew systems design. *Proceedings of the AGARD AMP Symposium on Situational Awareness in Aerospace Operations, CP478.* Seuilly-sur Seine, France: NATO AGARD.

196 Akbanov, M., Vassilios, G. and Logothetis, M.D., 2019, WannaCry ransomware: Analysis of infection, persistence, recovery prevention and propagation mechanisms, https://doi.org/10.26636/jtit.2019.130218.

197 National Cyber Security Centre (NCSC), 2021, Device Security Guidance, Guidance for organisations on how to choose, configure and use devices securely, Virtual Private Networks (VPNs), https://www.ncsc.gov.uk/collection/device-security-guidance/infrastructure/virtual-private-networks.

198 Pastor-Galindo, J., Nespoli, P. and Mármol, F.G., 2020, *The not yet exploited goldmine of OSINT: Opportunities, open challenges and future trends*, Special section of emerging approaches to cyber security, NJ: IEEE Access Vol. 8, pp. 10282–10304.

199 Ertran, A., Crossland G., Heath, C. and Jensen, R.J., 2018, *Everyday cyber security in organisations, literature review*, London: Royal Holloway University.

200 Helleman, N., 2023, Cyber-security and human psychology, *Security Awareness*, 25 Jul 2023, teiss magazine, https://www.teiss.co.uk/culture--people/cyber-security-and-human-psychology.

201 Microsoft, current in 2023, *Protect yourself from phishing*, https://support.microsoft.com/en-us/windows/protect-yourself-from-phishing-0c7ea947-ba98-3bd9-7184-430e1f860a44.

202 Lallie, H.S., Shepherd, L.A., Nurse, J.R.C., Erola, A., Epiphaniou, G., Maple, C. and Bellekens, X., 2021, Cyber security in the age of COVID-19: A timeline and analysis of cyber-crime and cyber-attacks during the pandemic, *Computers & Security* 105(2021), 102248.

203   Private Eye, March 2023, *Secret of Shame*, Stinky Phish, page 10.

204   Coatesworth, B. 2023, The psychology of social engineering. *Cyber Security: A Peer-Reviewed Journal* 6(3), 261–274.

205   Spar, B., Lefkowitz, R., Dye, C. and Pate, D., 2018, *2018 Workplace Learning Report*, LinkedIn Learning.

206   Ministry of Defence, 2023, JSP 822, Defence Direction and Guidance for Training and Education, Volume 2: Individual Training, version 6.

207   Witsenboer, J.W.A., Sijtsma, K. and Scheel, F., 2022, Measuring cyber secure behaviour of elementary and high school students in the Netherlands, *Computers and Education* 186(2022), 104536.

208   Biçakci, A. S. and Evren, A.G., 2022, Thinking multiculturally in the age of hybrid threats: Converging cyber and physical security in Akkuyu nuclear power plant, *Nuclear Engineering and Technology* 54(2022) 2467–2474.

209   Hollnagel, E., 2015, RAG – Introduction to the Resilience Analysis Grid (RAG), https://erikhollnagel.com/onewebmedia/RAG%20Outline%20V2.pdf.

210   David, D.P., Keupp, M.K. & Mermoud, A., 2020, Knowledge absorption for cyber-security: The role of human beliefs, *Computer in Human Behavior* 106(2020), 106255.

211   EU, 2023, *Directive on measures for a high common level of cybersecurity across the Union (NIS2 Directive)*, https://digital-strategy.ec.europa.eu/en/policies/nis2-directive.

212   Serrat, O., 2009, The five whys technique, *Knowledge Solutions*, February 2009, 30.

213   Ramsay, D.K., 2009, *The practical handbook of investigation* (2nd ed.), ISBN 1-897667-10-8.

214   Altaf, A., Thron, E., Faily, S., Dogan, H. and Mylonas, A., 2019, *Evaluating the impact of cyber security and safety within human factors in rail using attacker personas*, Aspect 2019.

215   Stanton, N.A., 2014, EAST: A method for investigating social, information and task networks, In S. Sharples and S. Shorrok (Eds.), *Contemporary ergonomics and human factors 2014*, Institute of Ergonomics and Human Factors, CRC Press (pubs).

216   Lane, B.R., Salmon, P.M., Stanton, N.A., Cherney, A. and Lacey, D., 2019, Using the Event Analysis of Systemic Teamwork (EAST) broken-links approach to understand vulnerabilities to disruption, Ergonomics, 62:9, 1134–1149, https://doi.org/10.1080/00140139.2019.1621392.

217   Health and Safety Executive (HSE), 2013, Managing for health and safety, HSG 65, ISBN 978 0 7176 6456 6.

218   Based on Doran, G. T. (1981). There's a S.M.A.R.T. way to write management's goals and objectives. *Management Review* 70(11), 35–36.

219   Argyris, C., 1977, Double-loop learning in organisations, *Harvard Business Review*, September-October 1977.

220   Aiyanyo, I.D., Samuel, H., Lim, H., 2020, A systematic review of defensive and offensive cybersecurity with machine learning, *Applied Sciences* 10, 5811.

221   Sayed, M.A., Anwar, A.H., Kiekintveld, C., Kamhoua, C., 2023, Honeypot allocation for cyber deception in dynamic tactical networks: A game theoretic approach. In Fu, J., Kroupa, T., and Hayel, Y. (Eds.). *Decision and Game Theory for Security*. GameSec 2023. Lecture Notes in Computer Science, vol. 14167. Cham: Springer. https://doi.org/10.1007/978-3-031-50670-3_10.

222 Smeets, M., 2018, The strategic promise of offensive cyber operations, *Strategic Studies Quarterly*, 12(3), 90–113.
223 Grant, T., Burke, I. and Van Heerden, R., 2012, Comparing models of offensive cyber operations, *Proceedings of 7th International Conference on Information Warface and Security* (ICIW 2012), Seattle WA.
224 Widowson et al., 2022, *Human Affected Cyber Security (HACS) Framework*, Chartered Institute of Ergonomics and Human Factors.
225 NIST, 2014, Framework for improving Critical Infrastructure Cyber security – Version 1.0, National Institute of Standards and Technology.

Practical materials to capture risk

7

PLANNING THE ASSESSMENT APPROACH

Previous chapters have discussed human-related cyber security vulnerabilities and solutions. This chapter draws on the theoretical framework from these chapters to provide practical materials for risk assessment. When planning an assessment, it is necessary to consider the group to be assessed. The smaller the survey group, the more representative the findings. Separate assessments for each department allow the analyst to compare results and prioritise corrective actions for higher risk teams. Figure 7.1 illustrates the assessment process.

Survey materials are designed to address the individual and organisational vulnerabilities discussed in previous chapters. A checklist was developed for security representatives, and a questionnaire was created to capture the views of a wider group of employees. The results from the checklist and questionnaire are analysed and mapped to maturity levels. Focus groups delve deeper into the causes of any poor scores. Relevant recommendations are implemented (e.g. as part of a transformation programme). Then the survey can be repeated to identify new risk scores and hopefully demonstrate a higher maturity level. Each stage of the process is described in this chapter.

The materials in this book are meant as a source of valuable information; however, they are not meant as a substitute for direct expert assistance. For this, the services of a competent human factors professional(s) should be sought.

DOI: 10.1201/9781003427681-7

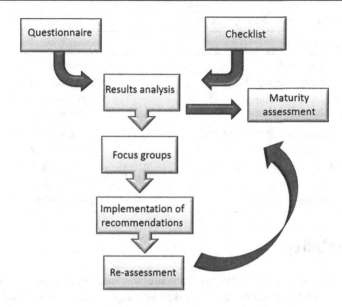

FIGURE 7.1   Assessment process.

QUESTIONNAIRE FOR EMPLOYEES

A Humans and Cyber Security (HaCS) questionnaire was designed to capture the perspective of a wide group of employees. One issue with questionnaires is that they are more likely to be completed by the most dutiful. As discussed in Chapter 4, conscientious, dutiful people are more likely to follow rules and procedures such as those pertaining to cyber security. Therefore, to minimise the risk of receiving a biased sample of responses, the completion time was kept to 10–20 minutes. Nevertheless, manager encouragement will be required to allow busy employees to allocate this time to complete the survey and therefore maximise the response rate. To manage expectations, a typical online questionnaire response rate is 44.1%,[226] so the sample size needs to reflect this for the results to be representative. Online calculators, like Raosoft,[227] can determine sample size. As a guide, for a population of 100, a sample of 80 is advisable.

TABLE 7.1   Test-retest ICC scores

CATEGORY	ICC
Organisational culture	0.980678233
Cyber security policy and procedures	0.851512739
Recruitment and staffing	0.99251737
Ergonomic design	0.97120857
Email, passwords, and the internet	0.942987025
Training and awareness	0.946191602
Physical working environment	0.991015981
Incident management	0.997185843
Malicious insider threat	0.995946802

Reliability

Test-retest reliability of the questionnaire was measured using a group of ten respondents. Each person was asked to complete the same questionnaire three times with intervals of at least a week between. The first two questionnaires were completed one week apart. The third repetition was completed eight weeks later. Eight of the respondents completed the same questionnaire three times as instructed. Their responses for each section of the questionnaire were analysed using the Intraclass Correlation Coefficient (ICC). This measure was selected because it calculates whether repeated measures of the same thing agree with each other.[228] The results are shown in Table 7.1.

As illustrated by the table, a high degree of inter-rater reliability was demonstrated in all sections of the questionnaire. It would be interesting to see if this pattern was repeated using a larger sample of respondents over more repetitions.

Internal consistency was tested using Cronbach's Alpha coefficient. Using a sample of 155 respondents, a score of 0.8654 was achieved, indicating a good level of internal consistency. In other words, the statistical results indicated that the questions all measured the same construct: cyber security. In summary, both tests indicated that the questionnaire was reliable.

Validity

Content validity was assessed by a group of eight experts using the Content Validity Index (CVI) by Waltz and Baussell (cited in Yaghmaie).[229] The index measured the relevance, clarity, simplicity, and ambiguity of each section of the questionnaire on a four-point scale. A score of 75% or above was considered appropriate. HaCS questionnaire raters were a mixture of three human factors, and four security experts plus one security culture specialist. According to the CVI, the number of raters who scored each item (question) highly (3 or 4) was divided by the total number of raters and multiplied by 100, to give the percentage score for each question. Two questions scored below this percentage (achieving 70%) so were modified. Specifically, more information was added to define cyber security before question #4 which addresses the perception of cyber security culture. The word 'security' was added to clarify the screening question (#7). The median score over all the questions was 100% in all categories. The mean score was 90% in all categories except for 'simplicity', which received an overall score of 100%. Good content validity was therefore demonstrated.

Testing

The questionnaire was initially tested on a group of six, comprising experienced human factors, systems engineering, and security professionals. It was updated after feedback from the group members. The questionnaire was then tested on a pilot group of 20 employees, and 19 responses were received. It was then tested on a much larger group (1,018). 'Microsoft Forms' was used as a platform for distributing the questionnaire and receiving results. Figure 7.2 depicts an extract from this questionnaire.

Other response options were accessed by using the horizontal scrollbar. Ideally, all options would be visible without the need to scroll. A link to the questionnaire was sent to participants via email.

Application

On receipt of the questionnaire, the following introductory statement is presented:

> This questionnaire is designed to measure cyber security in your organisation and identify vulnerabilities. It takes approximately 10 minutes to complete. The results will be used to enhance cyber security. Responses are anonymous.

On completion of the questionnaire, this outro statement is provided:

> Your response was submitted. Thank-you for your time. The collated anonymous responses will be used to enhance cyber security.

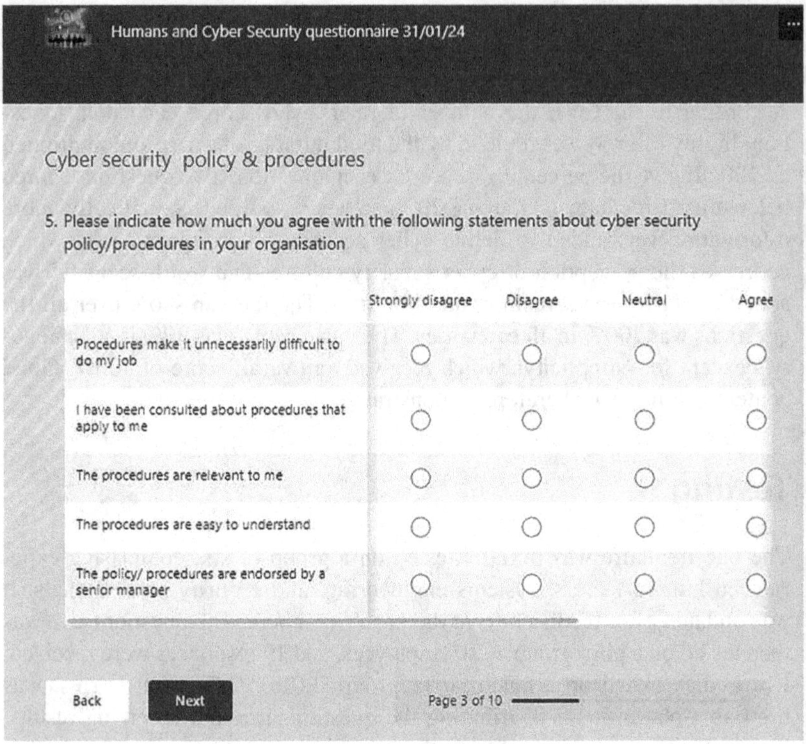

FIGURE 7.2   HaCS questionnaire extract.

The full, updated, questionnaire is shown in Table 7.2. The qualitative information in the first ('About you') section can be tailored according to the desired depth of analysis, for example by adding age, gender, and job grade. The length of time a respondent has worked for the organisation may affect responses such as perceptions of the culture. Knowledge of cyber security may affect training experiences. The other sections in the questionnaire and checklist correspond to the solutions in Chapter 6. Branch questions are included, where related questions only appear if the preceding question is answered accordingly. For example, if the answer to question 12 ("*Do you receive security reports*") is "no", the survey skips to question 14. If the answer is "yes", the respondent receives question 13, which asks about the reports. Note that the malicious insider threat section is named "Ethics and wellbeing" in the employee-facing questionnaire, to encourage honest responses. Response options and scoring are explained in the following section.

TABLE 7.2 HaCS questionnaire

Q#	ALL-EMPLOYEE QUESTIONNAIRE	RESPONSE OPTIONS	HIGH G/B	MAX SCORE	RECOMMENDATION #
About you					
1	Which department do you work for?	\<bespoke department names\>	N/A	N/A	N/A
2	How long have you worked for this organisation?	Less than 6 months, 6 months to 2 years, over 2 years	N/A	N/A	N/A
Cyber security experience					
3	How much knowledge do you have about cyber security?	Likert: None, very little, moderate, quite a lot, substantial	N/A	N/A	N/A
Organisational culture					
4	Cyber security is how we reduce the risk of cyber-attack. Cyber-attacks affect computer systems, networks, and devices. Please indicate how much you agree with each of the following statements				2CQ 2bICQ
a	My **managers** value cyber security.	Likert (agree) 'don't know' as middle option (score 'don't know' as 3)	G	5	
b	My **team** values cyber security.	"	G	5	
c	My managers value **productivity** over cyber security.	"	**B**	5	
d	Good cyber security behaviour is recognised/**rewarded** in my organisation.	"	G	5	

(Continued)

TABLE 7.2 (Continued)

Q#	ALL-EMPLOYEE QUESTIONNAIRE	RESPONSE OPTIONS	HIGH G/B	MAX SCORE	RECOMMENDATION #
Cyber security policy and procedures					
5	*Please indicate how much you agree with the following statements about cyber security policy and procedures in your organisation*				
a	Procedure(s) make it *unnecessarily* **difficult** to do my job.	Likert (agree) with 'don't know' on end	**B**	5	3aWCQ 3bWCQ 3cWCQ 3dWCQ
b	I have been **consulted** about security procedures that apply to me.	"	G	5	
c	The procedures are **relevant** to me.	"	G	5	3bWCQ 11bICQ
d	The procedures are easy to **understand**.	"	G	5	
e	The policy/procedure(s) are **endorsed** by a senior manager.	"	G	5	
Recruitment and staffing					
6	Do you deal with commercially and/ or **security sensitive (e.g. HMG classified) information** regularly as part of your job? *If yes:*	Yes, no			
7	How often have you undergone security **screening** for your current role?	Never, once, more than once	G	3	4cICQ 4dIC

No.	Statement	Response options	Type	Scale	Code
8	Have you taken a **personality test** in support of your role?	Never, once, more than once	G	3	4cICQ 4dICQ

Third parties (heading not shown on form)

No.	Statement	Response options	Type	Scale	Code
9	Does your organisation mandate security requirements in **supplier contracts**?	Yes, don't know, no	G	3	5ICQ

Ergonomic design

Sub-title: Human Factors (HF) or ergonomics is a scientific discipline that uses knowledge of human mental and physical strengths and limitations to design products and systems around the people that use them. HF professionals are accredited by the CIEHF, BPS, HFES, IEA[a] or an equivalent professional body. UX design seeks to optimise the user experience.

No.	Statement	Response options	Type	Scale	Code
10	Equipment Human Factors (HF) and User experience (UX)				
a	**Competent** personnel have measured the usability of software and/or hardware I use.	Likert (frequency) 'don't know' on end	G	5	6aWCQ
b	HF/UX requirements are considered when **purchasing** equipment.	Likert (frequency) 'don't know' on end	G	5	6dWCQ
11	Ergonomic design of equipment/applications you use: *Please indicate how much you agree with each of the following statements*				6bWQC 6cWCQ
a	My work **software** applications are easy to use.	Likert (agree) with N/A	G	5 (don't score N/A)	
b	My work **hardware** equipment is easy to use.	Likert (agree) with N/A	G	5	
c	I have enough **time** to view and respond to the alarms I receive.	Likert (agree) with N/A	G	5	

(Continued)

TABLE 7.2 (Continued)

Q#	ALL-EMPLOYEE QUESTIONNAIRE	RESPONSE OPTIONS	HIGH GIB	MAX SCORE	RECOMMENDATION #
12	Do you receive cyber security **reports**? *If yes:*	Yes, no			7WCQ
13	Cyber security reports you receive:				
a	Are easy to understand	Likert (agree)	G	5	
b	Contain the right amount of information for you	Likert (agree)	G	5	
Email, passwords, and the internet					
14	*Please indicate the frequency of these activities pertaining to passwords*				10CQ
a	Are you able to choose your own work **password(s)**?	Likert (frequency)	G	5	10dCQ 10aWCQ
b	Do you use a **password manager** tool?	Likert (frequency)	G	5	10bWCQ
c	I use the same password for more than one website/application (e.g. for work and home).	Likert (frequency) 'don't know' on end	B	5	10aWCQ 10bWCQ
d	My colleagues share login information with each other.	" "	B	5	10aWCQ 10bWCQ
e	I use a Personal Electronic Device (PED) (e.g. mobile phone, tablet, laptop, USB) for work emails and/ or documents.	" "	B	5	9WCQ
f	I store my password(s) in a PED	" "	B	5	9WCQ

15		*Please indicate how much you agree with this statement about* **password recall**				
	a	It is easy is it to remember my password(s).	Likert (agree)	G	5	10aWCQ 10bWCQ
16		I have been made aware of restrictions to sharing information on **video-conferencing** platforms.	Likert (agree) 'N/A' on end	G	5	10eiiiWCQ
17		*Please indicate the frequency of the following activities pertaining to* **emails**				
	a	I have enough time to manage emails I receive at work.	Likert (frequency)	G	5	10diiC 10diiiCQ 9WCQ
	b	I use my personal email account for work.	" "	B	5	10iiWCQ
	c	I use free public Wi-Fi for work.	" "	B	5	2CQ
	d	I receive emails from within my organisation inviting me to click a link.	" "	B	5	
18		Social media e.g. *LinkedIn. Please indicate the frequency of the following activities.*				
	a	My colleagues discuss work on social media.	Likert (frequency) 'don't know' on end	B	5	10eivWCQ
	b	I access location tracking applications (e.g. for fitness) in the office/working area.	Likert (frequency) 'don't know' on end	B	5	11lWCQ
19		I know how to protect my personal and work information on social media (e.g. *LinkedIn*).	Likert (agree)	G	5	10eivWCQ 11CQ

(Continued)

TABLE 7.2 (Continued)

Q#	ALL-EMPLOYEE QUESTIONNAIRE	RESPONSE OPTIONS	HIGH GIB	MAX SCORE	RECOMMENDATION #
Training and awareness					
20	How often have you completed training about cyber security at your current organisation? *If never, go to Q21, otherwise go to Q22.*	Likert (frequency), 'don't know' on end	G	5	11CQ 11piiCQ
21	Have you ever received training about cyber security (e.g. at a previous organisation)? *If never, go to Q24. Otherwise, go to Q22:*	Likert (frequency), 'don't know' on end	G	5	—
22	*Please indicate how much you agree with each statement about the* **quality** *of cyber security training you received*				
a	The training was **relevant** to me.	Likert (agree)	G	5	11eWCQ 11cWCQ
b	I found the training **informative**.	" "	G	5	11CQ
c	The training changed my attitude towards cyber security.	" "	G	5	11aICQ
d	The training was **engaging**.	" "	G	5	11fWCQ
e	The training was easy to follow.	" "	G	5	11CQ
23	Training **content** - *Have you received training about*				
a	How do cognitive biases and decision-making affect cyber security?	Likert (frequency)	G	5	11dWCQ
b	Phishing email detection?	Likert (frequency)	G	5	11hWCQ

c	Security risks associated with **AI**?	Likert (frequency)	G	5	11jWCQ
d	Cyber security risks associated with your **physical working environment**?	Likert (frequency)	G	5	11lWCQ
e	Passwords use?	Likert (frequency)	G	5	11mWCQ
f	Social engineering tactics?	Likert (frequency)	G	5	11nWCQ
g	Cyber security incident management?	Likert (frequency)	G	5	11gWCQ
24	Has your cyber security knowledge been **assessed** (e.g. in a test as part of training)?	Likert (frequency)	G	5	11qlCQ
25	Do you know how data protection legislation (e.g. **GDPR**[2]) affects your work?	No, slightly aware, moderately aware, mostly aware, fully aware	G	5	11kWCQ
26	*Please indicate how much you agree with these additional statements about cyber security **training***				
a	I have enough time to complete cyber security training	Likert (agree)	G	5	11o(i&ii)WCQ
b	I receive too many test phishing emails from my security team.	Likert (agree)	B	5	11iWCQ
c	I am confident I would be able to identify a cyber security attack.	Likert (agree)	G	5	11qlCQ
27	Who has the most **responsibility** for cyber security?	IT/Information security department; My manager; Me; All equally responsible			11bWCQ

(Continued)

TABLE 7.2 (Continued)

Q#	ALL-EMPLOYEE QUESTIONNAIRE	RESPONSE OPTIONS	HIGH G/B	MAX SCORE	RECOMMENDATION #
					12CQ
Physical working environment					
28	*Please indicate the frequency of these activities*	Likert (frequency) 'N/A' on end			
a	People hold **doors** open for others without checking their access credentials.		B	5	12bWCQ 12cWCQ
b	I lock my **computer** when I leave my desk.	" "	G	5	12dWCQ 12eWCQ
c	I lock away sensitive **paperwork**.	" "	G	5	12fWCQ
d	I **carry** sensitive paperwork away from the work site.	" "	B	5	12fWCQ
e	I see paperwork left on **printer(s)**.	" "	B	5	12gWCQ
f	The bin for **disposal** of sensitive hard copy information is emptied before it overflows	" "	G	5	12hWCQ
g	The **clear desk/whiteboard** policy is followed.	" "	G	5	12dWCQ 12fWQQ
h	I can find somewhere to discuss sensitive work without being **overheard.**	" "	G	5	12iiCQ 12jWCQ
i	People **wear** organisation-branded clothing/lanyards in public.	" "	B	5	12kWCQ
j	I have heard people mention the organisation's name when **talking about work** in public.	" "	B	5	12kWCQ

k	**Visitors** are booked in advance.	" "	G	5	12lWCQ
l	I see people I **don't recognise** in my work environment.	" "	**B**	5	12lWCQ
m	People are **challenged** if they don't wear ID on organisation premises.	" "	G	5	12lWCQ

29 *Please indicate how much you **agree** with these statements about the physical working environment*

a	I would feel awkward about requesting someone's credentials before allowing them through a door to a secure area.	Likert (agree)	**B**	5	12bWCQ
b	It is easy to identify when visitors are present in my work environment.	Likert (agree)	G	5	12lWCQ
c	It is easy to identify whether a visitor requires an escort.	Likert (agree)	G	5	12lWCQ

Incident management 13CQ

30 *Please indicate the **frequency** of these activities about cyber security incident management*

a	I would report a cyber security incident and/or suspicious behaviour if I saw it.	Likert (frequency)	G	5	13aiWCQ 13aiiWCQ 13ciiWCQ
b	I receive information about cyber security attacks on my organisation.	Likert (frequency)	G	5	13aiWCQ 13ciiCQ
c	I have supported a cyber security incident investigation.	Likert (frequency)	G	5	13d1WCQ

(Continued)

TABLE 7.2 (Continued)

Q#	ALL-EMPLOYEE QUESTIONNAIRE	RESPONSE OPTIONS	HIGH G/B	MAX SCORE	RECOMMENDATION #
31	Please indicate how much you **agree** with these statements about cyber security incident management				
a	I would be worried about being blamed if I reported an incident.	Likert (agree) N/A on end	**B**	5	2CQ
b	I know how to report an incident in my organisation.	Likert (agree) N/A on end	G	5	13aiiWCQ
c	I find it easy to report an incident/suspicious behaviour.	Likert (agree) N/A on end	G	5	13aiiWCQ
d	Actions/lessons from cyber security incident investigations are reviewed at the onset of projects.	Likert (agree) N/A on end	G	5	13eii WCQ 13eIC
Ethics and wellbeing ('Malicious insider threat' - not included in form)					14ICQ
32	Please indicate how much you agree with these statements				
a	I generally support the key management decisions in my organisation.	Likert (agree)	G	5	2CQ
b	My employer is socially and ethically responsible.	" "	G	5	2CQ
c	My organisation values diversity.	" "	G	5	2CQ
d	My manager and colleagues respect my beliefs.	" "	G	5	2CQ
e	My manager is supportive towards me.	" "	G	5	2CQ
f	I have access to emotional support if I need it.	" "	G	5	2cWCQ

g	I am rewarded fairly for effort I put into my work.	" "	G	5	2CQ
h	I have no serious concerns in my personal life.	" "	G	5	14ICQ
i	I need more excitement in my work.	" "	B	5	14ICQ
j	I'm worried about my job security.	" "	B	5	14ICQ
k	I'm worried about my personal financial situation.	" "	B	5	14ICQ
33	Please provide any further comments about this questionnaire and/or cyber security in your organisation.	Free text			N/A

a BPS – British Psychological Society; CIEHF - Chartered Institute of Ergonomics and Human Factors; HFES - Human Factors and Ergonomics Society; IEA - International Ergonomics Association; USB – Universal Serial Bus.

QUESTIONNAIRE SCORING

Response options are provided for each question. Unless otherwise specified, Likert questions address frequency (*Never, Rarely, Sometimes, Often, Always*) or agreement (*Strongly disagree, Disagree, Neutral, Agree, Strongly agree*). The numerical score for each response is provided in Table 7.3. All responses in a category use the same scale (unless they are not included in the quantitative scoring).

In most cases, the higher the score, the stronger the cyber security. These questions are depicted by a 'G' (for good cyber security). However, some questions are reversed. These exceptions are indicated by the letter 'B' (for bad cyber security) in bold text, in a shaded cell in Table 7.2. An example of default and reversed scoring is shown in Table 7.4.

TABLE 7.3   Questionnaire scoring

RESPONSE	*SCORE*
Likert	1–5
Yes	2
No	1
Once	2
More than once	3
Slightly aware	2
Moderately aware	3
Mostly aware	4
Fully aware	5
Don't know	Do not score (except Q4, where the score is 3 and Q9, where the score is 2)
N/A	Do not score

TABLE 7.4   Likert frequency scoring

Default score	1	2	3	4	5	Don't score
Response	Never	Rarely	Sometimes	Often	Always	N/A
Reversed score	5	4	3	2	1	Don't score
Response	Never	Rarely	Sometimes	Often	Always	N/A

$$\frac{\bar{x}\ received\ score}{\bar{x}\ maximum\ score} \times 100$$

FIGURE 7.3   Section percentage score calculation.

The responses of all the respondents can be averaged per question. The questionnaire comprises the following nine sections, which correspond to the sections in the solutions chapter:

- About you
- Organisational culture
- Cyber security policy and procedures
- Recruitment and staffing
- Ergonomic design
- Email, passwords, and the internet
- Training and awareness
- Physical working environment
- Incident management
- Malicious insider threat (termed "Ethics & wellbeing" in the employee-facing questionnaire).

The percentage of good cyber security can be calculated for each section. First, the average maximum *possible* score per section is calculated using the maximum possible score for each question in the section, shown in the fifth column of Table 7.2. This can be used to divide the average *actual* score received per section. The result is then multiplied by 100 to derive a percentage, as illustrated in Figure 7.3.

The final column in the table depicts the number(s) of the recommendation(s) designed to address the subject of each question. Recommendations are defined in Table 7.7.

CHECKLIST FOR SECURITY SPECIALISTS (STRUCTURED INTERVIEW TEMPLATE)

Whereas the questionnaire was designed to capture the perspective of employees and completion time was kept relatively short, the checklist considers specialist knowledge and requires more time to complete (20–30 minutes if all the information is known). It can be populated directly by security personnel or serve as a template for a structured interview conducted by human factors professionals. Depending on the knowledge and remit of the organisation's security personnel, it may be necessary for additional personnel (such as recruitment and site facilities representatives) to complete the sections on recruitment and physical site security, respectively.

The checklist sections correspond to those in the questionnaire and in Chapter 6. Some of the questions (numbers 4a, 6b, 8a & b, 10a, 11a & d, 15a, 18, 20a, 31, 33, 34a & c-k, 35, 36c, 37, 40, 48, 52, 53a & c, 58, 59b, 60a, 70h & i), tackle similar issues to those in the questionnaire but from a different perspective. A small number of questions (numbers 6c, 12a & b, and 57), are identical, to enable the checklist and questionnaire results for those questions to be compared.

Reliability

A test sample of two people independently completed the same checklist on the same area. Because the checklist is designed for security specialists, the pool of suitable respondents was smaller than the one used to test the all-employee questionnaire. The Spearman's Correlation Coefficient score was 0.737, indicating that both raters agreed, and the checklist is reliable. However, this sample is small, and further testing would be required to investigate statistical significance.

Checklist scoring

Like the questionnaire scoring, unless otherwise specified, Likert questions address frequency (*Never, Rarely, Sometimes, Often, Always*); or agreement (*Strongly disagree, Disagree, Neutral, Agree, Strongly agree*). The numerical score for each checklist item is provided in Table 7.6. In default cases, the higher the score, the stronger the cyber security. Reversed options are depicted by the letter 'B' in bold text, in a shaded cell in Table 7.5.

All the questions within a section use the same rating scale for consistency. The cyber security percentage score can be calculated for each section, based on the *actual* average score received divided by the *maximum* possible average score (indicated in bold text, in a shaded square next to the section name in Table 7.5), as illustrated in Figure 7.3. An example section percentage calculation is included in the scoring extract example shown in Figure 7.4.

The final column in the table depicts the number(s) of the recommendation(s) designed to address the subject of each question. Recommendations are defined in Table 7.7.

TABLE 7.5 HaCS checklist for security specialists

Q#	CHECKLIST/INTERVIEW FOR SECURITY PERSONNEL	RESPONSE OPTIONS	HIGH G/B	MAX SCORE	RECOMMENDATION #
Risk and maturity assessment.					
Please indicate the frequency of each of these activities					
				5	**1C**
1	Cyber security **risk** assessments	Likert (frequency)	G	5	1aIC
2	Organisational **maturity** assessments	" "	G	5	1aIC
3	**Repeat** risk assessments, after recommendations have been implemented	" "	G	5	1bIC
4	**Organisational culture**			**2**	**2CQ**
a	Have you conducted a **transformation** programme to enhance cyber security? *If yes:*	Yes, no, don't know, N/A	G	2	2CQ
b	Does the transformation programme consider **culture types**?	" "	G	2	2aIC 2bIC
c	Do you have a mechanism for reporting employee **wellbeing** issues?	" "	G	2	2cWC 2eWC
d	Are all managers advised of their **duty of care** to their staff?	" "	G	2	2eWC
e	Do you actively encourage cyber security behaviours as part of objective setting and performance **appraisals**?	" "	G	2	2dWC
f	Do you actively support cyber security training in **schools**?	" "	G	2	2fIC

(Continued)

TABLE 7.5 (Continued)

Q#	CHECKLIST/INTERVIEW FOR SECURITY PERSONNEL	RESPONSE OPTIONS	HIGH G/B	MAX SCORE	RECOMMENDATION #
Policy and procedures				**5**	
5	Do we have a cyber security policy(ies)?	Likert (agree) 'don't know' on end	G	5	3CQ
6	*Please indicate the frequency of each activity pertaining to cyber security policy and procedures*				
a	Are employees **consulted** about the security procedures they are required to adopt?	Likert (agree) 'don't know' on end	G	5	3CQ
b	Are security procedures specifically designed to avoid employee **workarounds?**	" "	G	5	3a WCQ
c	Are the cyber security policy/procedure(s) **endorsed** by a senior manager?	" "	G	5	3d WCQ
Recruitment and staffing				**5**	
7	Please indicate CISO's level of influence in the organisation.	Likert (Very high (C-suite); high; moderate; low; very low)	G	5	4aWC
8	Are potential employees **screened** when they apply for roles with access to sensitive information? *If 'never', go to Q9, otherwise go to Q8a.*	Likert (frequency)	G	5	4cICQ

a	Do security candidates undergo **personality** testing?	Likert (frequency)	G	5	4cICQ
b	Do security candidates have **credit and blackmail** checks?	Likert (frequency)	G	5	4cICQ
9	*Please indicate the extent to which you agree with each statement about security staffing*				
a	We have **enough** competent cyber security personnel.	Likert (agree)	G	5	4bIC
b	There is clear **accountability** for cyber security between the IT and security teams.	" "	G	5	4dIC
c	Firewalls are implemented where necessary to protect information from **contractors and suppliers**.	" "	G	5	5ICQ
10	*Please indicate the frequency of each activity pertaining staffing and job design*				
a	Cyber security requirements are included in **supplier** contracts.	Likert (frequency)	G	5	5ICQ
b	Cyber security requirements are included in customer contracts.	" "	G	5	5ICQ
c	Contractors and suppliers are provided with **training** in the organisation's cyber security policy and procedures.	" "	G	5	5ICQ
d	Information is **restricted** to minimise the amount any individual can access.	" "	G	5	4dIC

(Continued)

TABLE 7.5 (Continued)

Q#	CHECKLIST/INTERVIEW FOR SECURITY PERSONNEL	RESPONSE OPTIONS	HIGH G/B	MAX SCORE	RECOMMENDATION #
	Ergonomic design			**5**	6CQ
	Human Factors (HF) or ergonomics is a scientific discipline that uses knowledge of human mental and physical strengths and limitations to design products and systems around the people that use them. HF professionals are accredited by the CIEHF, BPS, HFES, IEA or an equivalent professional body. UX design seeks to optimise the User eXperience.				
11	*Please indicate the frequency of each of the following activities*				
a	Have **competent personnel** measured the usability or user experience (UX) of your software and/or hardware?	Likert (frequency) 'don't know' on end	G	5	6aWCQ 6bWCQ
b	Does security software necessitate a large reliance on the user's **memory capacity?** (i.e. do you have to hold a lot of elements in your head?)	" "	**B**	5	6aWCQ 6bWCQ
c	Have you conducted **user-testing** (e.g. applying SART (Situational Awareness Rating Technique); SUS (System Usability Scale); workload tools; open-ended questions)?	" "	G	5	6cWCQ
d	Do you consider human factors/UX requirements when **purchasing** equipment?	" "	G	5	6dWCQ

12	Cyber security reports you **receive**:				
a	Are easy to understand	Likert (agree) N/A on end	G	5	7WCQ
b	Contain the right amount of information for you	" "	G	5	7WCQ
13	Do you consider HF presentation of information requirements for the target audience when **producing** security reports?	Yes, no	G	5	7WCQ
14	*The following questions are about portable equipment*				
a	Do you provide **work phones** with security controls?	Likert (frequency)	G	5	9WCQ
b	Are employees advised not to use Personal Electronic Devices **(PEDs)** for work?	" "	G	5	9WCQ
c	Do you monitor transfer of information using **USBs** and personal email accounts?	" "	G	5	9WCQ
Email, passwords and the internet					
Passwords				**2**	
15	Is a password **policy** in place? *If yes:*	Yes, no	G	2	10CQ
					10aWCQ
					10bWCQ
15a	Are employees allowed to **choose** their own password(s) in accordance with the policy?	Yes, no	G	2	10aWCQ
16	How often are password changes **enforced**? (*choose the nearest answer*)	Monthly or more; less frequently than monthly	G	2	10aWCQ

(Continued)

TABLE 7.5 (Continued)

Q#	CHECKLIST/INTERVIEW FOR SECURITY PERSONNEL	RESPONSE OPTIONS	HIGH G/B	MAX SCORE	RECOMMENDATION #
17	Have you observed passwords **written** near the devices they are intended to protect?	Yes, no	**B**	2	10aWCQ 10bWCQ
18	Do you offer a **password manager** to employees? *If yes:*	Yes, no	G	2	10bWCQ
18a	Do you make them aware of it and provide instructions in use?	Yes, no	G	2	10bWCQ
19	Have you considered alternatives to passwords for user validation of organisation devices (e.g. **biometrics**)?	Yes, no, don't know	G	2	10cIC
Emails					
20	*Please indicate the frequency of these activities pertaining to emails*			**2**	10dCQ
a	Managers encourage employees to allow enough **time** to manage their emails.	Yes, no	G	2	10diiWCQ
b	Key members of staff with access to sensitive information have administrative **support** to help them manage their emails.	" "	G	2	10diiiWC
21	Are emails marked to indicate when they are from an **external** source?	" "	G	2	10dvWC
22	Is **technology** in place to filter out malicious emails before they reach the employee recipient?	" "	G	2	10divIC

#	Question	Options			Code
23	Do you provide alternative **communication media** to reduce reliance on email applications?	"	G	2	10diiCQ
Internet					
24	*Please indicate the frequency of these activities pertaining to internet use*			**2**	**10ecQ**
a	Software **downloads** are restricted by administrator access.	Yes, no	G	2	10eiWC
b	Unauthorised software downloads are **monitored**.	Yes, no	G	2	10eiWC
c	**OSINT** (Open Source Intelligence) is conducted.	Yes, no	G	2	10eiWWCQ
d	**Personal details** about family/friends are included on organisation websites.	Yes, no, N/A	B	2	10eiWWCQ
25	Is there an **access log** for important sensitive systems?	Yes, no	G	2	8IC / 10eIC
26	Is access to malicious websites **blocked** where known?	Yes, no	G	2	10eiWC
27	Are appropriate sites/companies **whitelisted**?	Yes, no	G	2	10eiWC
28	Is access to any **social media** sites blocked from work devices?	Yes, no	G	2	10eiWC
29	Is **anti-virus** protection implemented?	Yes, no	G	2	10divIC
30	Is the use of a **VPN** enforced?	Yes, no	G	2	10eiiWCQ
31	Do you monitor when **public Wi-Fi** is used?	Yes, no, N/A	G	2	10eiiWCQ

(Continued)

TABLE 7.5 (Continued)

Q#	CHECKLIST/INTERVIEW FOR SECURITY PERSONNEL	RESPONSE OPTIONS	HIGH G/B	MAX SCORE	RECOMMENDATION #
Tele/video-conferencing					
32	Are tele/video-conferencing platforms **reviewed** to determine the level of security information that can be shared?	Yes, no	G	2	10eiiiWCQ
33	Are users made aware of the **security restrictions** on the information they can share on specific tele/video-conferencing platforms?	Yes, no	G	2	10eiiiWCQ
Training and awareness				**5**	
34	*Training content - Have you provided advice/training about*				
a	How employees can protect their personal and work information on **social media**?	Likert (frequency)	G	5	11CQ
b	**Location tracking** applications?	" "	G	5	11lWCQ
c	Cognitive **bias** and risky decision-making?	" "	G	5	11dWCQ
d	**AI** risks (e.g. deepfakes, AI-generated phishing emails, generative AI IP loss risk)?	" "	G	5	11jWCQ
e	**Data protection** legislation?	" "	G	5	11kWCQ
f	Risks associated with the **physical working environment** (e.g. tailgating tactics, dangers of public Wi-Fi, USB use, unlocked paperwork, and computers)?	" "	G	5	11lWCQ
g	The design and use of **passwords**?	" "	G	5	11mWCQ

h	**Incident response** (to relevant personnel)?	" "	G	5	11gWCQ
i	**Realistic examples** of incidents and vulnerabilities, recent, relevant to the target audience?	" "	G	5	11eWCQ
j	**Social engineering** tactics	" "	G	5	11nWCQ
k	**Phishing** email detection?	" "	G	5	11hWCQ
35	How often are phishing email **simulations** conducted?	" "	G	5	11iWCQ
36	*Training design*				
a	Is the training specifically designed to capture **intrinsic motivation** of employees?	Likert (frequency)	G	5	11aIC
b	Does the training consider what appeals to different **personality** types?	" "	G	5	11cWCQ
c	Are employees encouraged to take **responsibility** for their own cyber security?	" "	G	5	11bICQ
d	Do you collect **feedback** on the quality of the training?	" "	G	5	11q
	Sufficient time for training				
37	Do (senior) managers allow their staff the **time** to complete cyber security training?	Likert (frequency)	G	5	11oiWCQ
38	Do employees have **flexibility** about when they take the training (to allow them to fit it around their other work commitments)?	" "	G	5	11oiiWCQ
39	Is training provided as part of **onboarding**?	" "	G	5	11piiICQ

(Continued)

TABLE 7.5 (Continued)

Q#	CHECKLIST/INTERVIEW FOR SECURITY PERSONNEL	RESPONSE OPTIONS	HIGH G/B	MAX SCORE	RECOMMENDATION #
40	Is refresher training provided at regular intervals **throughout employment?**	" "	G	5	11piiCQ
41	Has a Training Needs Analysis (**TNA**) been conducted to determine the frequency of training?	" "	G	5	11piiCQ
42	*Training measurement - please indicate how much you agree with the following statements*				
a	All, or most, employees have successfully **completed** the training.	Likert (agree)	G	5	11qICQ
b	We know the level of **understanding** of each trainee.	Likert (agree)	G	5	11qICQ
Physical working environment					
43	Do the **physical and cyber security** teams communicate regularly?	Yes, no	G	2	12CQ
				2	12aWC
44	Do you have **turnstiles** and/or doors requiring identification on key entry points?	" "	G	2	12bWCQ
45	Do you have a **security official** at entry points to secure areas (even if intermittently)?	" "	G	2	12bWCQ
46	Do you have **notices on doors** to secure areas advising employees to (a) show their pass when someone holds the door open and (b) ask to see access credentials when holding the door open for someone else?	" "	G	2	12bWCQ

47	Do you **share a building** with another organisation(s)?	""	B	2	12bWCQ 12iiICQ
48	Are **privacy filters** for monitors provided?	""	G	2	12eWCQ
49	Is there a process for management of confidential **documents**?	""	G	2	12hWCQ
50	**Server rooms** are usually locked when not occupied by authorised personnel.	""	G	2	12cWCQ
51	Regular **patrols** are conducted to police a clear desk/whiteboard policy and check for locked computers and paperwork left on printers.	""	G	2	12dWCQ
52	Employees generally follow the **clear desk/ whiteboard and printer** policy.	""	G	2	12dWCQ
53	*These questions are about sensitive conversations*				
a	Employees are warned about identifying the organisation and **discussing work in public.**	""	G	2	12jWCQ 12kWCQ
b	Employees are encouraged to **intervene** if they hear sensitive information being discussed in public.	""	G	2	12jWCQ 12kWCQ
c	The work environment includes enough spaces where employees can have **sensitive conversations.**	""	G	2	12iiICQ
54	Most employees **value** physical security controls.	""	G	2	2CQ

(Continued)

TABLE 7.5 (Continued)

Q#	CHECKLIST/INTERVIEW FOR SECURITY PERSONNEL	RESPONSE OPTIONS	HIGH G/B	MAX SCORE	RECOMMENDATION #
Visitors					
55	Is there a visitor management **policy**? *If yes:*	Yes, no	G	2	12IWCQ
56	What is the **average number** of visitors compared to staff members to the workplace per month (calculated as the number of staff members divided by the number of visitors)?	0–4.99; more than 5; not recorded	G	2	12IWCQ
57	Visitors are booked in **advance**	Yes, no	G	2	12IWCQ
58	Visitors are provided with ID to allow employees to **distinguish close-up and from a distance**, who is a visitor and whether they need to be escorted by a member of the organisation.	Yes, no	G	2	12IWCQ
Incident management					
59	*Sharing incidents*			**2**	13CQ
a	Are incidents shared with the **board**?	Yes, no	G	2	13aiWCQ 13aiiWCQ
b	Are incidents shared with all **employees**?	Yes, no	G	2	13aiWCQ
c	Are significant incidents shared **externally**, with **other organisations**?	Yes, no	G	2	13aiiiWC
60	*Incident reporting*				
a	Do you have a system for **reporting** incidents and suspicious behaviour? *If yes:*	Yes, no	G	2	13aiiWCQ

b	Is the incident reporting system **anonymous**?	" "	G	2	13aiiWCQ
c	Do you record the **number of people who report** cyber security incidents? *If yes:*	" "	G	2	13ciiWC
d	Do you record **how quickly** people report cyber security incidents?	" "	G	2	13ciiWC
e	Do you have an incident **response** procedure?	" "	G	2	13bIC
f	Do you have an incident **recovery** procedure?	" "	G	2	13bIC
g	Do you have a **SOC** (Security Operations Centre)?	" "	G	2	13bIC
h	Is an **incident database** maintained to capture up-to-date internal and external threats and vulnerabilities, incidents and near misses?	" "	G	2	13ciICQ
Asset management					
61	A comprehensive asset management system is in place and up to date.	Yes, no	G	2	8IC
62	Do you have a procedure for reporting and measuring lost/stolen devices?	Yes, no	G	2	13ciiiWCQ
63	Do you know how many laptops and other work devices were lost/stolen in the last year?	Yes, no, N/A	G	2	13ciiiWCQ
64	Are laptops encrypted?	Yes, no, N/A	G	2	9WCQ
65	Are work phones and tablets encrypted?	Yes, no, N/A	G	2	9WCQ

(Continued)

TABLE 7.5 (Continued)

Q#	CHECKLIST/INTERVIEW FOR SECURITY PERSONNEL	RESPONSE OPTIONS	HIGH G/B	MAX SCORE	RECOMMENDATION #
66	Software **patches** are up to date.	Yes, no	G	2	8IC
67	The **operating system** is updated according to security requirements.	" "	G	2	8IC
68	**Redundant accounts** (e.g. from leavers) are closed and access rights are removed in a timely manner.	" "	G	2	8IC
Incident investigation					
69	*Are incidents investigated? If yes:*				13dC
70	*The following questions are about incident investigation*				
a	A **multidisciplinary team** is used to investigate significant incidents	Yes, no	G	2	13diWC
b	A competent **human factors** specialist is routinely part of incident investigation teams	Yes, no	G	2	13diWC
d	**Root cause analysis** methodology is applied, and/or an incident architecture is created.	Yes, no	G	2	13diiWC
e	A broad **range** of contributory factors are considered.	Yes, no	G	2	13diiiWC
g	**Learning teams** are employed.	Yes, no	G	2	13eIC

h	The **review of SMART actions** from incident investigations is actively encouraged at the onset of projects and when setting up teams.	Yes, no	G	2	13eiWC 13eiiWCQ
i	Employees are **reprimanded** when they are found to have contributed to security incident.	Yes, no	B	5	13aiiWCQ

Malicious insider threat

71	*Please indicate the frequency of these activities*		**5**	**5**	**14ICQ**
a	**Managers** report suspicious employee behaviour.	Likert (frequency)	G	5	14ICQ
b	**Employees** report suspicious employee behaviour.	Likert (frequency)	G	5	14ICQ
c	**HR representatives** report suspicious employee behaviour.	Likert (frequency)	G	5	14ICQ
d	Employee **morale** is checked (e.g. with employee engagement surveys).	Likert (frequency)	G	5	2CQ
72	Are **'high-risk'** individuals monitored?	Likert (frequency)	G	5	14ICQ
73	Are **system anomalies** monitored?	Likert (frequency)	G	5	14ICQ

TABLE 7.6   Checklist scoring

RESPONSE	SCORE
Likert	1–5
Yes	2
No	1
Never	1
Once	2
More than once	3
Monthly or more	2
Less frequently than monthly	1
0–4.99	1
More than 5	2
Don't know	Do not score
N/A	Do not score
Not recorded	Do not score

Q#	Checklist/interview for security personnel	Scores	Average score per section	% good cyber security	Numerical score
	Risk and maturity assessment. *Please indicate the frequency each of these activities are conducted*	Risk & maturity assessment	2.7	0.5	
1	Cyber security risk assessments	Sometimes			3
2	Organisational maturity assessments	Sometimes			3
3	Repeat risk assessments, after recommendations have been implemented	Never Rarely			2
4	Organisational culture	Sometimes Often	1.5	0.8	
a	Have you conducted a transformation programme to enhance cybersecurity?	Frequently			2

FIGURE 7.4   Checklist extract showing scoring.

RESULTS ANALYSIS (QUESTIONNAIRE AND CHECKLIST)

'Microsoft Forms' automatically generates averaged results for each question in the survey. A representative example from a test group is shown in Figure 7.5.

The average score for each of the nine sections can be presented graphically, as illustrated by Figure 7.6, for a quick understanding of the areas of highest risk.

The questionnaire and checklist results for the average section scores can be calculated separately and/or combined. Of particular interest for

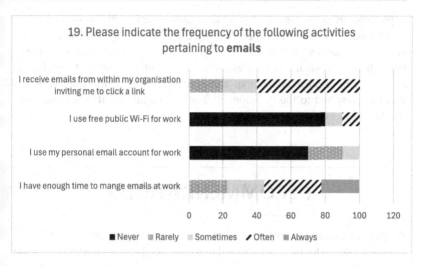

FIGURE 7.5   Questionnaire results extract.

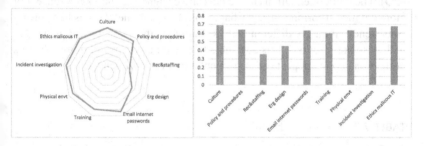

FIGURE 7.6   Percentage of good cyber security radar diagram and bar chart.

comparison may be the results of questions that are duplicated in both survey tools. The scores can be used to identify the organisation's cyber security maturity level (see later). Once the results have been collated and analysed, focus groups can delve into the causes of the highest risk scores.

Focus groups

Small groups of six to eight people, representing different types of employees, can be consulted regarding the survey results, as well as any security policy workarounds. This technique may form part of an organisational transformation programme. Recommendations associated with each question in the survey represent the desired state. The focus groups can work to identify how to progress from the current, "as is" state (demonstrated by the survey) to the desired one. Unfortunately, people can be tempted to respond to questionnaires in a way that makes them look good.[159] Focus groups allow people to voice their interpretation of undesirable results without casting themselves in a negative light. Participants can also reveal non-verbal information by the way they respond to the questions.[230] The technique can also help alleviate the dutiful response bias described earlier.

RECOMMENDATIONS

Recommendations are presented in Table 7.7. They correspond to the solutions in Chapter 6. Codes have been added to recommendation numbers to indicate 'quick wins' (W) versus 'in-depth' solutions (I), and whether the recommendation pertains to questions from the questionnaire (Q), checklist (C), or both (CQ). The recommendations in the table are designed for quick reference. For more detailed information, see Chapter 6. The recommendations help ensure consistency amongst different raters, although they can be supplemented with bespoke solutions. It is advisable to consult a competent human factors professional(s) to support the implementation of recommendations.

TABLE 7.7 Recommendations

RECOMMENDATION #	CODE	RECOMMENDATION DESCRIPTION
1	**C**	**Conduct risk, vulnerability, and maturity assessment.**
1a	IC	Conduct risk and vulnerability assessment to identify 'as is' maturity level.
1b	IC	Retest risk assessment after recommendations have been implemented (*from this checklist/Chapter 6, to identify desired goals for transformation programme*) and determine a new level of maturity.
2	**CQ**	**Assess and transform organisational culture**

(Continued)

TABLE 7.7 (Continued)

RECOMMENDATION #	CODE	RECOMMENDATION DESCRIPTION
2a	IC	Identify organisational culture types
2b	ICQ	Implement a culture transformation programme (based on culture type(s) (Table 6.1 has several sub-recommendations e.g. recognise good cyber security behaviours).
2c	WCQ	Provide emotional support.
2d	WC	Encourage cyber security behaviours as part of objective setting and performance appraisals.
2e	WC	Advise managers of their duty of care to their staff.
2f	IC	Support cyber security training in school as soon as children go online.
3	**CQ**	**Design cyber security policy and procedures around users.**
3a	WCQ	Investigate workarounds. Encourage employees to challenge processes based on their experience. Identify what intrinsically motivates the employees.
3b	WCQ	Encourage ownership of cyber security through participatory design. *(See also recommendation 11b - deter overreliance)*
3c	WCQ	Make policies accessible (easy to find, containing consistent terminology, not onerous).
3d	WCQ	Senior managers to endorse policy and procedures
4	**CQ**	**Consider HF in recruitment and staffing.**
4a	WC	Ensure the CISO reports to the board/CEO.
4b	IC	Ensure there are enough security personnel. (Develop people if required).
4c	ICQ	Screen candidates for roles with access to sensitive information. (Conduct personality testing for key roles, credit checks, and other blackmail checks via Developed Vetting (DV)).

(Continued)

TABLE 7.7   (Continued)

RECOMMENDATION #	CODE	RECOMMENDATION DESCRIPTION
4d	ICQ	Design jobs to minimise access to sensitive information. Ensure clear accountability between IT and security teams.
5	ICQ	Manage third parties (suppliers, contractors, customers): Incorporate cyber security requirements into contracts; provide training in policy/procedures; and create firewalls to restrict access to information.
6	**CQ**	**Ergonomic design**
6a	WCQ	Consider HF in equipment design (including task and workload analysis, anthropometry, biomechanics, controls, viewing angle/distance, and HRA). Use competent HF personnel
6b	WCQ	Apply HF to HCI design. Considerations include alarms management; pattern recognition; feedback; colour, auditory, and blink coding; and Neilsen's heuristics.
6c	WCQ	Conduct user-testing, for example using SART, SUS, workload tools, and/or open-ended questioning.
6d	WCQ	Consider HF in the procurement and selection of equipment.
7	WCQ	Apply HF to report presentation: Consider the presentation of information and colour; tailor for the target audience - not too complex for non-security personnel, avoid jargon.
8	IC	Manage assets: Record status; update anti-virus protection; conduct regular, timely software patching; and update operating systems according to security requirements.
9	WCQ	Discourage use of PEDs for work (including USBs) and provide work phones, laptops, tablets, etc., with security controls/encryption.
10	**CQ**	**Manage email, passwords and the internet**

(Continued)

TABLE 7.7   (Continued)

RECOMMENDATION #	CODE	RECOMMENDATION DESCRIPTION
10a	WCQ	Allow users to choose their passwords.
10b	WCQ	Provide password managers/ safes (with instructions).
10c	ICQ	Investigate alternatives to passwords (e.g. biometrics).
10d	*CQ*	*Manage email*
10di	ICQ	Reduce the number of emails received by considering alternative means of communication.
10dii	WCQ	Encourage people to allow more time to manage emails.
10diii	WC	Provide administrative support to manage emails for key personnel.
10div	IC	Invest in technological solutions to protect systems and filter out malicious emails.
10dv	WC	Mark emails to indicate when they are from an external source.
10e	*ICQ*	*Manage internet usage*
10ei	WC	Restrict software download access rights. Whitelist appropriate websites and companies to minimise the risk of workarounds.
10eii	WCQ	Use a VPN
10eiii	WCQ	Manage tele/video-conferencing
10eiv	WC	Conduct OSINT.
11	**CQ**	**Provide (quality) cyber security training and awareness** (including online behaviour and social media).
11a	ICQ	Capture motivation in training: Identify what intrinsically motivates the employees/ target audience, including personality differences.
11b	ICQ	Deter over-reliance on others such as the IS department.
11c	WCQ	Tailor training to the trainee's personality.

(Continued)

TABLE 7.7   (Continued)

RECOMMENDATION #	CODE	RECOMMENDATION DESCRIPTION
11d	WCQ	Include cognitive bias and risky decision-making awareness in training.
11e	WCQ	Incorporate realistic examples, recent, relevant to the target audience (not alarmist) in training. Include the dangers of pop-up messages as people have been found to spend little time reading them.
11f	WCQ	Make training engaging (e.g. using gamification, regular nudges, clothing bearing the word "intruder").
11g	WCQ	Provide incident response training. Content could include role-play; cognitive biases e.g. cognitive narrowing (dealing with the immediate threat but not bigger picture); making decisions with insufficient information versus deliberation; impulsiveness; risky decision-making (e.g. avoid shutting internet operations down because of certain loss, despite the risk of greater loss); allow all to voice their concerns.
11h	WCQ	Provide training in phishing email detection.
11i	WCQ	Conduct phishing simulations (in moderation and in conjunction with other solutions).
11j	WCQ	Provide training in AI risks.
11k	WCQ	Provide data protection legislation training.
11l	WCQ	Provide training in physical environment security. Include attacker tailgating tactics; dangers of public Wi-Fi; unlocked paperwork computers; location tracking; and USB use.
11m	WCQ	Provide training in password management.
11n	WCQ	Provide training in social engineering tactics
11o	CQ	*Allocate sufficient time for training*
11oi	WCQ	Senior managers to encourage employees to block out time for learning.
11oii	WCQ	Make training flexible in design to fit around busy schedules.
11p	C	*Manage frequency and timing of training*

(Continued)

TABLE 7.7   (Continued)

RECOMMENDATION #	CODE	RECOMMENDATION DESCRIPTION
11pi	ICQ	Conduct training at induction and throughout employment e.g. with regular prompts of smaller training with tests. Possibly conduct TNA to determine frequency based on complexity and frequency the learning is applied.
11q	ICQ	Create and maintain a competence management system
12	**CQ**	**Secure the physical working environment**
12a	WCQ	Ensure joint communication between physical and cyber security teams.
12b	WCQ	Secure entry points and server rooms, for example, by physical pen-testing; turnstiles; intermittent security presence; and notices on doors).
12c	WC	Lock server rooms.
12d	WCQ	Practice and monitor good housekeeping: Implement a clear desk policy; lock computers when leaving desks; close blinds and switch off lights in visible areas at night; regularly inspect housekeeping.
12e	WCQ	Provide privacy filters for monitors.
12f	WCQ	Provide lockable storage for sensitive paperwork (including when in transit).
12g	WCQ	Prevent remote printing.
12h	WCQ	Dispose of sensitive documents.
12i	ICQ	Provide secure spaces for sensitive conversations.
12j	WCQ	Where lack of privacy, restrict information discussed and avoid mention of the company name.
12k	WCQ	Discourage the wearing of company lanyards and identifiable clothing in public. Where not possible e.g. when marketing, advise staff to be cautious about the information they discuss.

(Continued)

TABLE 7.7   (Continued)

RECOMMENDATION #	CODE	RECOMMENDATION DESCRIPTION
12l	WCQ	Manage visitors: book them in advance; require ID (badges should make it obvious who is a visitor and whether they need escorting); and encourage people to challenge others if they are not wearing ID.
13	**CQ**	**Manage incidents**
13ai	WCQ	Share incidents internally, with the board and employees.
13aii	WCQ	Facilitate incident reporting: Make it easy and quick (e.g. link on email application and front page of intranet); reward – do not punish; use a confidential reporting system.
13aiii	WC	Share incidents externally.
13b	IC	Respond to incidents. *(also see 11g – incident response training)*
13c	C	*Measure incidents*
13ci	ICQ	Create and maintain an incident database, containing internal and external threats and vulnerabilities.
13cii	WCQ	Record the number of people who report incidents and how quickly.
13ciii	WCQ	Identify the number of lost/stolen laptops and other devices.
13d	C	*Investigate incidents*
13di	WC	Invite a multidisciplinary team of stakeholders, including competent HF specialists.
13dii	WC	Conduct root cause analysis/create an incident architecture.
13diii	WC	Consider the organisational and individual vulnerabilities in this book (avoid recommending training as only solution).
13e	IC	Learn from incidents
13ei	WC	Generate SMART actions from incident investigations.
13eii	WCQ	*Implement* SMART actions from investigations and review at onset of projects and when setting up teams.
14	ICQ	Manage malicious behaviour.

TABLE 7.8   Example GRADE maturity scoring

Maturity level		Percentage score	Culture	Policy & procedures	Recruitment & staffing	Ergonomic design	Email internet & passwords	Training	Physical envt	Incident investigation	Ethics & wellbeing (malicious IT)	Overall average percentage	
Enhanced	5	81-100											5
Developed	4	61-80	4	4			4		4	4	4		4
Alert	3	41-60						3				3	3
Reactive	2	21-40			2	2							2
Growing	1	0-20											1

MATURITY ASSESSMENT APPLICATION

Maturity assessment is described in Chapter 6. The questionnaire and check-list results represent a level of cyber security maturity. The percentage scores (per section and overall) can be mapped to one of five maturity levels in the GRADE framework: Growing (1), Reactive (2), Alert (3), Developed (4), or Enhanced (5), as illustrated in Table 7.8. In this hypothetical example, recruitment & staffing, and ergonomic design are the least mature and therefore need the most attention. Application of recommendations from Chapter 6 (summarised in Table 7.7) will enhance the scores.

Re-assessment

After relevant recommendations have been implemented to reduce risk and enhance cyber security maturity, the survey can be repeated. The new results should demonstrate higher levels of maturity. The re-assessment may be done for individual departments or for the overall organisation. It can be repeated at regular intervals to monitor risks and maturity and identify priority areas for investment. The GRADE scores can be used as a dashboard to monitor and report organisational maturity over time.

Chapter 8 describes findings from applied surveys.

NOTES

226 Wu, M.-J., Zhao, K., Fils-Aime, F., 2022, Response rates of online surveys in published research: A meta-analysis, *Computers in Human Behavior Reports* 7, 100206, ISSN 2451-9588, https://doi.org/10.1016/j.chbr.2022.100206.

227 Raosoft_®, 2004, Random size calculator, retrieved February 2024 from http://www.raosoft.com/samplesize.html.
228 Bridger, R, 2017, *Introduction to human factors and ergonomics* (4th ed.). CRC Press, Taylor and Francis Group (pubs), Boca Raton FL. ISBN: 9781498795944.
229 Yaghmaie, F. 2003, Content validity and its estimation, *Journal of Medical Education* 3(1), 25–27.
230 Citizens Advice, 2015, How to run focus groups. Solutions for equality and growth, Equality & diversity forum, co-funded by the European Union, National Association of Citizens Advice Bureaux, registered charity number 279057.

Applied examples of risk identification and mitigation

8

This chapter aims to capture the applied experience of conducting human factor assessments of cyber security. For security reasons, case-study examples have been redacted, although mitigation strategies have been put in place to address vulnerabilities identified at the time. Early assessments used the Cyber Human Error Assessment (CHEAT®) tool.[83] More recently, the HaCS questionnaire and checklist, described in Chapter 7, were applied. Multiple organisations were included in the analysis.

DATA COLLECTION SUMMARY

In the nine case studies assessed, the data was collected via a questionnaire and semi-structured interviews. In some cases, workshops or focus groups were also conducted. Data collection methods and logistics were discussed in advance with an information security point of contact from the relevant organisations. Considerations included identification of the target departments or groups; the method of questionnaire distribution; questionnaire recipients, interviewees, and their contact details; and the method of fore-warning employees about the survey. Table 8.1 summarises the number of completed questionnaires returned, questionnaire response rate, and the number and duration of interviews, for each case study (where the information was available).

All questionnaires were distributed by a contact who was internal to the respective organisation and known to the employees. Two weeks was the time

DOI: 10.1201/9781003427681-8

TABLE 8.1   Case study data collection

CASE-STUDY #	NUMBER OF QUESTIONNAIRES RETURNED	QUESTIONNAIRE RESPONSE RATE (%)	NUMBER OF INTERVIEWS	INTERVIEW DURATION (MINUTES)
1	15	100	5	30–60
2	37	45	8	30–60
3	7	14	6	30–60
4	90	34	10	30–75
5	63	80 (estimated)	13	30–75
6	6	33	6	30–75
7	57	82	11	30
8	63	85	8	30
9	19	95	2	20

usually allowed for completion. Questionnaire completion reminders were issued in all cases. Arguably, the response rate could be said to be indicative of the cyber security culture. Subsequent implementation of prioritised recommendations enhanced cyber security. Common findings from these assessments are described in the next section.

COMMON FINDINGS IN HUMAN FACTORS CYBER VULNERABILITY INVESTIGATIONS

A thematic analysis was conducted to determine common findings from nine organisational assessments of the human element in cyber security. Ten common themes were identified, as shown in Table 8.2. They are presented in order of impact.

Interestingly, risky human behaviours relating to the physical working environment were common across all case studies analysed. Vulnerabilities of this nature allow attackers to breach the secure perimeter and access sensitive hard copy and electronic materials and networks. All the vulnerabilities and associated solutions are discussed earlier in this book. A proactive risk assessment approach is beneficial in highlighting vulnerabilities like these, to generate risk-reduction strategies.

TABLE 8.2 Thematic analysis of assessment findings

PERCENTAGE OF ASSESSMENTS AFFECTED %	THEME	DESCRIPTION
100	Physical working environment	Tailgating; unlocked doors; lack of clear desk policy
78	Ergonomics	Poor design/lack of ergonomics consideration
67	Training	Lack of awareness of cyber security threats/ vulnerabilities; Lack of competence management
56	Time pressure/shortage of IS personnel	Work under time pressure; Lack of competent Information Security (IS) personnel
56	Workarounds	*(see next section)*
44	Suppliers, contractors, customers	Lack of security consideration/policy
44	Email management	Lack of time to deal with the number of emails received
33	Lack of reward	Good cyber security behaviours were not rewarded
22	Passwords	A lot to remember
22	Procedures	Lack of consultation

Example workarounds

If work systems are not designed around the people who use them, they will likely find workarounds or shortcuts. It is human nature to try to find the easiest way to perform a task. If left unchecked, workarounds and poor behaviours can become the norm. In the case studies analysed, specific workarounds, which could be detrimental to organisational resilience, were identified through interviews.

In one organisation, pass-controlled doors were sabotaged. Rubber was placed over magnetic strips to negate the need for a security pass. It would have been difficult to identify the culprit(s) and punishment would have been counterproductive. Instead, the investigation into the cause of the workaround suggested that the security door was considered to be an inconvenience, and personnel did not fully appreciate the risks. This is an example of a cultural issue. Where the practice had been allowed to continue, some employees had apparently assumed that others considered it to be acceptable. Awareness training for all staff in the vicinity is recommended to prevent this type of workaround.

In other organisations, security doors were left open, or keys were left out for others to use. Ease of access to keys may save time by removing the need to find the key-holder or operate a key-safe. However, it obviously eliminates the security benefits of the lock. Access devices (such as keys or secure smartcards) can be issued to authorised individuals to remove the need to leave keys near the doors they open.

Investigations in one organisation identified that doors were left ajar because the door entry code system was noisy and would wake people who were sleeping nearby at night. In other words, the poor security behaviour stemmed from a social concern, rather than malicious motives. In this situation, it may be tempting to simply instruct personnel to lock the door. However, the best solution would be to replace the security entry mechanism for a quieter one, to eliminate the perceived need to keep the door open.

In some of the organisations, the sharing of user login information was relatively commonplace. This behaviour makes it easy for people to access information for which they are not authorised. Workarounds like this can be caused by a desire for a seamless handover between shifts; the need to access information quickly; difficulty in obtaining credentials; overly complex access requirements; or simple laziness. In any case, the causes need to be investigated with the personnel concerned so that bespoke solutions (such as technology or shift pattern design) can be identified and implemented.

Another common workaround was the use of Personnel Electronic Devices (PED), and personal software applications, for work. In one organisation, it was discovered that PEDs were used to order spare parts, risking exposure of equipment status to unauthorised factions. Ergonomic/UX design of applications would facilitate efficient part requests and thereby reduce the risk of this sort of workaround. In one organisation, a popular social application was used to send images of items of concern. If the messages had been intercepted, the knowledge of these organisational vulnerabilities could have been exploited by malicious actors. People find popular social applications such as this desirable because of their usability. Alternative tools therefore need to be equally usable. Ergonomics/UX can help with that too.

Some organisations ban PEDs in certain areas, for example, because of the photographic and recording capabilities of modern devices. Assessors found there were PEDs in a prohibited area. If deliberate, this suggests the employees wanted, or needed, a continuous means of communication with external parties. A more secure means would reduce the temptation to breach this security requirement. Of course, sometimes people may forget to leave their PEDs outside the restricted area. In those cases, reminders can be placed near entry points, and security patrols can police the policy to reduce the risk.

Some employees justified the use of PEDs because it was difficult to access the internet on work devices. Dubious websites were accessed in this way. They felt that the information they were dealing with was already available to the public, so did not see the point of the security restrictions. This highlights a cultural norm issue, where poor performance (in this case a leak of sensitive information) was accepted, and the need to protect information was neglected. Culture transformation programmes and awareness training can be beneficial here.

Personal email accounts were used by employees, in more than one organisation, to share work information. This means the information was not protected by the employers' security systems. Further scrutiny revealed that the work application prevented them from sharing large files with customers and suppliers. Time pressure exacerbated the need to share information quickly. Discussions were opened between the relevant employees and IS personnel. This paved the way for a safer working practice.

These examples indicate the need for organisations to identify workarounds. This can be achieved through focus groups and one-to-one interviews. In many cases, the risk of workarounds can be reduced by consulting employees about the procedures they are required to adopt and finding the optimal balance between security and usability.

HACS CHECKLIST APPLICATION

The HaCS checklist described in Chapter 7 was applied by two raters. The small sample reflects the relative shortage of security specialists, for which it was designed, compared to the wider employee population. An Excel spreadsheet was used to capture the responses and calculate average scores for the categories. The scores of both raters were averaged.

Overall, good cyber security was indicated, with the physical environment, incident management, and internet protection scoring particularly highly (90% secure). Development areas identified are withheld for security reasons. Checklist scores can be interpreted in conjunction with questionnaire scores to capture a wider range of opinions.

LESSONS LEARNT FROM PRACTICAL ASSESSMENTS

Experience of applied assessments of the human element in cyber security yielded lessons that were applied to later investigations.

Allow adequate time for interviews

One of the first findings was the tendency to underestimate the amount of time needed to arrange and conduct interviews. Interviewees were hard to get hold of and did not always prioritise time for the activity. Financial compensation for taking part may work for some, but many interviewees have other commitments and would be unlikely to be swayed by such a strategy. Instead, it is important to allow a long enough time period to find diary space for the interviews.

Build rapport with interviewees

Once in an interview, it is necessary to build a rapport. Bear in mind what the interviewee might be anticipating. Some people interviewed had obviously felt the need to prepare, as if they were being tested on their cyber security knowledge. Whilst the information they provide can be very useful, the interview can become overly scripted in these cases. The interviewer needs to try to get the interviewee to relax in order to obtain an honest picture of the culture of cyber security. Sometimes, however, it does not work. Some people just do not want to be drawn out. Another good lesson, therefore, is knowing when to stop. If an interviewee remains resistant, it may be time to move on to the next question or even the next person.

Be prepared to steer the interview

Other people are very happy to chat and view the interview as a welcome break from their work. The trick here is knowing when and how to steer the conversation, to ensure the key security questions can be covered within the available time.

Make friends with the internal questionnaire administrator

Survey planning activities are typically conducted in consultation with a CISO and/or other security personnel. Indeed, management endorsement can increase the response rate. However, the distribution of questionnaires may be delegated to someone else, and that person is likely to have other work demands. Therefore, it is important to ensure that the questionnaire administrator is adequately informed about the process and its priority.

It can be hard to control the number of questionnaires distributed by an internal contact, so regular communication with the administrator is necessary to understand who has received the survey and to determine the response rate and the level of representativeness of the data.

Send questionnaire reminders

Questionnaire response rate can be hard to achieve. It was usually necessary to remind people to complete the questionnaire after the initial two-week deadline. These reminders typically resulted in a new wave of submitted responses. Regular promotion of the survey can be helpful too.

Supplement data collection with documentation

Documents such as job descriptions and policies can be useful in addressing gaps in interview and questionnaire data. Researching these documents can reduce interview time and compensate when some interviewees are unavailable in the required timescale.

Sometimes unanticipated factors, outside the control of the assessors, may occur. However, preparation and cognisance of lessons learnt can prevent or reduce the impact of these occurrences, maximising the benefits of the assessment.

Summary and conclusions

9

Cyber security is an ongoing challenge. It affects significant aspects of our lives, from our personal information to national infrastructure and competitive industry. Many reports attribute cyber security incidents to "human error", as discussed in Chapter 2. Multiple attack vectors and examples are described, all with a human element. The discipline of human factors can help us understand human behaviour in the context of cyber security. More importantly, it provides us with solutions to mitigate risks in that context.

A lot of research focuses on measuring culture and changing behaviour. The implication is that human behaviour is at fault. Routinely blaming the user is strongly criticised by human factors and safety culture experts. Instead, it is argued that wider factors such as organisation, management, training, technology, environment, and previous events are also at play. Indeed, humans can be an asset, as well as a threat, to cyber security, demonstrating flexibility, situational awareness, and decision-making capabilities that technology traditionally has not. This book therefore addresses organisational factors and human ones.

Organisational considerations in Chapter 5 include culture type and morale; design and accessibility of cyber security procedures; recruitment and reporting chain; third parties (such as the supply chain and customers); training and awareness; technology design, usability, and misuse; the physical working environment and incident management. Training is not the only answer.

There are a variety of theories about human motivation and behaviour. Chapter 3 discusses several of these and summarises them into six categories, depicted by the TOMS model. Behaviour motivators in the model include fear of negative consequences; coping efficacy; anticipation of reward; intrinsic desires; autonomy over behaviour; social norms; and basic needs. However, people are arguably more complex than that, so it is also advisable to consider individual differences. Chapter 4 addresses such factors, including personality, attitudes and behaviour, age, and gender, and their impact on cyber security. Decision-making style can also affect cyber security behaviour. This chapter discusses the impact of specific cognitive biases and mental

DOI: 10.1201/9781003427681-9

shortcuts, as well as risk-taking behaviours under different circumstances. Error is classified to help analysts understand the causes of behaviour in incidents. Error-producing conditions, such as workload and time pressure, are discussed in terms of their relevance to cyber security.

Solutions to address both individual human behaviour and organisational issues are described in Chapter 6. Tools to measure the risks associated with these vulnerabilities are discussed in Chapter 7. A checklist for security personnel as well as an all-employee questionnaire are provided. Incorporated questions to test each vulnerability are associated with a recommendation(s) from Chapter 8. After the data from tools such as these has been collected and analysed, focus groups can probe the reasons behind the findings. The assessment tools have been tested, but additional research and repeated application are encouraged to develop the instruments and knowledge base in this area. Analysis by competent human factors professionals will enhance the quality of the assessments. To find competent professionals, contact the Chartered Institute of Ergonomics and Human Factors (CIEHF) or International Ergonomics Association (IEA) and associated bodies.

Results from the data collection and analysis can be used to assess the overall maturity of the organisation in terms of human factors and cyber security. The 'GRADE' framework captures five levels of maturity for each category in the questionnaire and checklist. Maturity may be upgraded after key prioritised recommendations have been successfully implemented and the organisation has been re-assessed. GRADE scores can be used as a dashboard to monitor and report organisational maturity over time.

Finally, case-study examples of human factors in cyber security assessments are outlined in Chapter 8. Common findings from assessments were identified and categorised. Several identified workarounds are discussed. Practical lessons derived from the investigations are provided for the benefit of future assessments of human factors in cyber security.

Index

Printed in the United States
by Baker & Taylor Publisher Services